The City

The City in the Developing World

Robert B. Potter

and

Sally Lloyd-Evans

 Longman

Addison Wesley Longman Limited
Edinburgh Gate
Harlow
Essex CM20 2JE
United Kingdom
and Associated Companies throughout the world

First published 1998

ISBN   0 582 35741 1

Visit Addison Wesley Longman on the world-wide web at
http://www.awl-he.com

British Library Cataloguing in Publication Data

A catalogue record for this book is available from the British Library

Library of Congress Cataloging in Publication Data

A catalog record for this book is available from the Library of Congress

Set by 35 in 11/12pt Adobe Garamond
Produced by Addison Wesley Longman Singapore (Pte) Ltd.,
Printed in Singapore

For

Sam

and

Katherine

Contents

Contents

List of figures

List of plates

List of tables

Preface

This textbook has been designed as an introduction to the study of towns and cities in developing countries, for undergraduate and Masters level candidates taking courses in institutions of higher education. We were drawn to the direct and forceful nature of the title *The City in the Developing World*. However, we should stress at the outset that use of the definite article as the first word in the title is intended to emphasise the salience of the role which urban settlements play in developing areas. Use of the expression 'The City' is not intended to imply that we believe that there is any such thing as a typical or archetypal urban form in the developing world. Indeed, one of the major themes which is explored in the text is the diversity of urban circumstances which currently appear to be resulting from so-called 'global processes'.

As well as cherished themes which students need to consider fully, such as the progress of world urbanisation, the nature of urban primacy and the provision of low-income housing, we have endeavoured to stress topics of pressing contemporary relevance such as the relations between cities and social surpluses, globalisation, convergence, divergence, modernity and postmodernity, poverty alleviation, education and health, emerging urban forms, child and female labour, environment and urbanisation, sustainability, empowerment, governance and so-called 'new urban management programmes'. Gender issues receive particular attention in many of the chapters, rather than being dealt with as a separate topic. We hope that the outcome is a balanced account incorporating both established and emerging concepts and ideas. For example, the process of globalisation is first discussed in relation to the structure of urban systems, in an account which endeavours to link together established notions concerning urban primacy with newer perspectives on global convergence, divergence, global shifts, and the emergence of global or world cities. The concept of globalisation then features in later chapters when considering employment, work and the city, and in relation to evolving urban structure. Dealing with such contemporary phenomena and processes, we have been fortunate in working with a publisher that respected our view that this type of textbook needs to be well illustrated, with both photographic plates and line diagrams.

We are, of course, grateful to many people, not least our colleagues and those whom we have taught, in helping to shape and sharpen our ideas. In addition, we should particularly like to thank Sophie Bowlby, Erlet Cater, Dennis Conway, David Hilling, Cathy McIlwaine, Michael Lloyd-Evans,

Virginia Potter, Andrew Powell and David Simon, all of whom read and commented on various draft chapters. Heather Browning assisted with some of the cartography, whilst Erika Meller prepared the photographic prints from colour slides. Kathy Roberts helped with typing up the tables. We should also like to record our appreciation of all who have been involved with the preparation and production of this volume at Addison Wesley Longman, for their manifest care and efficiency.

We hope that this concise introduction will act as a useful starting point for those teaching and learning about urban phenomena and urban processes in the developing world. We look forward to receiving the reactions of teachers and students in the form of reviews and direct communications via our respective institutional addresses. It is also our fervent hope that the content of the constituent chapters of this book will encourage undergraduates and postgraduates alike to undertake thesis and dissertation work set in developing countries.

Robert Potter
Sally Lloyd-Evans
November 1997

Acknowledgements

We are grateful to the following for permission to reproduce copyright material:

Croom Helm for figure 1.1 from R.B. Potter's *Urbanisation and Planning in the Third World: Spatial Perceptions and Public Participation* (Croom Helm: New York: St Martins Press, 1985); and Oxford University Press for figures 1.2, 1.3 and 2.2 from R.B. Potter's *Urbanisation in the Third World* (OUP: Oxford, 1995).

Thanks also to Janet Carlsson for plate 7.4 and Astrid Bishop for plate 8.2.

Introduction

This book has been written and designed specifically as an up-to-date and comprehensive introduction to the fascinating and policy-relevant field of urbanisation in developing countries. The primary target audience is undergraduates in the fields of geography, development studies, planning, economics and the social sciences defined more generally, who are either taking specialist courses on urban processes in the Third World, or reading for options which deal with development studies, development theory, gender and development or issues of Third World development more generally.

As there are already some very good books dealing with urbanisation, urban processes and cities in developing countries, it is clearly important to chart what we see as the distinctive aims and objectives of this textbook. As authors, our overarching goal has been to place an understanding of the developing world city in its wider global context. Firstly, this is done by means of developing the concept of **social surplus product** as a key to understanding the character of the contemporary Third World city. Secondly, throughout this text, the city in developing areas is centrally placed in the context of **global social, economic, political and cultural change**. Thus, the important themes of **globalisation** and **postmodernity** are examined both in relation to the structure of sets of towns and cities which make up the national or regional urban system, and in respect of ideas and concepts dealing with the morphology, structure and social patterning of individual urban areas. In this latter context, the focus is on how global processes are affecting the individuals and households which make up cities, and how people are responding on a day-to-day basis. Here characteristics such as age, gender, class and ethnicity are an important focus. Another major aim of the book is to illustrate and exemplify the argument that all theories of development can be viewed as explicit theories of urbanisation.

Taken together, these aims are closely reflected in the structure of the book. Chapters 1 and 2 are introductory in the sense of defining the nature and scope of urbanisation in the Third World, viewing this in broad historical terms since the advent of cities and urban living, and thereby developing the concept of social surplus product as a central theme for analysis in the remainder of the book.

Broadly speaking, the book is divided into three parts. Chapters 1 to 4 look at the issues surrounding urban systems and change in developing areas of the world. Equally broadly defined, Chapters 5–8 then focus attention on what is happening within cities in Third World regions. The last two chapters (9 and 10) are devoted to the consideration of policy issues, involving urban environmental conditions and questions of urban environmental sustainability, plus the wider arena of urban policies and strategies.

In overall terms, the book endeavours to explore in detail the relationships which exist between cities in developing areas and attendant processes of globalisation and change. An overarching argument is that global processes are not leading to uniformity between cities in different parts of the world. Far from it: although the generic problems faced by cities in terms of housing deficits, congestion and unemployment show considerable commonalities, urban processes seem to be leading to further differentiation between regions and places. In many instances, trends towards further localisation can be identified as the direct outcome on the ground. In this respect the processes of **global convergence** and **global divergence** act as overarching concepts informing our analysis of the city in the developing world.

As well as examining such vital topics as housing, employment and work, topics which are generally covered in the majority of texts on urbanisation in developing areas, the present volume emphasises issues which up to this point have customarily received less attention, or which have been almost entirely neglected. These include urbanisation and the provision of basic needs, including education, health, food and nutrition, the links between urban systems structure and development, the morphology and structure of cities in the developing world, and cities and the concept of environmental sustainability.

Another aim of the volume is to link the analysis of the Third World city with current themes of debate in the social sciences. It has already been noted that these include globalisation, modernity–postmodernity, convergence, divergence, the significance of gender and ethnicity, especially in relation to housing, basic needs and employment, world cities, global cities, the influences of structural adjustment policies and neo-liberalism, the new poverty agenda, the salience of civil society, along with the potential influence of new social movements.

The final chapters of the book make the telling point that it seems more than likely that environmental factors will be of pressing significance in shaping the future of cities in the Third World. The policy agenda, which is currently increasingly driven by the World Bank's 'urban push', and what is described as the new urban management programme, is the subject of discussion in the very final chapter of the volume.

Chapter 1

The nature and scale of urbanisation in the developing world

Introduction

One of the most frequently cited statistics summarising the process of urbanisation which is currently being experienced in the so-called developing world is that the towns and cities of these poorer countries are receiving a staggering 45 million new urban inhabitants each and every year. This vast number of new city dwellers in the poorer countries of the world – amounting to somewhere in the region of 125,000 new urban citizens a day worldwide – is the outcome of high levels of rural-to-urban migration in combination with high rates of natural increase of the population. By comparison, approximately 7 million urban residents are added on an annual basis to the towns and cities in the countries of the more developed world.

The scale of this process of urbanisation is difficult to comprehend in respect of the numbers of houses, water connections, schools, clinics, hospital beds and jobs that will be required over the coming decades in the more impoverished countries of the world. Quite simply, we are living through what can only be described as a record-breaking era: one during which the world has experienced its fastest ever rate of urbanisation. The level of urbanisation pertaining to a region, nation or any other territory is measured as the proportion of the total population that is to be found living in towns and cities, however these are defined, but normally following local definitions. Between 1950 and 2025, a period of some 75 years, the overall level of world urbanisation will have increased from 29 to 61 per cent. The half-way point at which 50 per cent of the world's population is to be found living in urban places is set to be passed shortly after the year 2000. Between 1960 and 1970, the world's urban population grew by 16.8 per cent. From 1970 to 1980, it increased again by 16.9 per cent. If the same rate were to have continued from 1980 onward, the world would have been totally urbanised by the year 2031, a period of just over 50 years. Such is the magnitude of the urban processes which is to be faced as we enter the twenty-first century.

These few illustrative statistics demonstrate that urbanisation and urban growth are occurring much more rapidly in the developing world than they did in the more developed world regions during the heyday of the Industrial Revolution. Later in this introductory chapter, statistics showing the rate of urbanisation in the various continental regions making up both the more developed and less developed world between now and the first quarter of the twenty-first century will be considered in detail. But before that, the changing history of world urbanisation which has just been sketched out in the broadest of terms, will be amplified more fully.

The magnitude of the developmental pressures presented by the current global urban process is perhaps put in more accessible terms by a hypothetical scenario which was presented in the news magazine *Newsweek*, under the admittedly very alarmist title 'An age of nightmare cities: flood tides of humanity will create mammoth urban problems for the Third World':

> It is a sweltering afternoon in the year 2000, in the biggest city ever seen on earth. Twenty-eight million people swarm about an 8-mile-wide mass of smoky slums, surrounding walled-in, high-rise islands of power and wealth. Half the city's work force is unemployed, most of the rich have fled and many of the poor have never even seen downtown. In a nameless, open-sewer shanty-town, the victims of yet another cholera epidemic are dying slowly, without any medical attention. Across the town, the water truck fails to arrive for the third straight day; police move in with tear gas to quell one more desultory riot. And at a score of gritty plazas around the city, groaning buses from the parched countryside empty a thousand more hungry peasants into what they think is their city of hope. (*Newsweek*, 31 October 1993, p. 26)

Historical perspectives on world urbanisation

As noted by the Brandt Report (1980) *North–South: a Programme for Survival*, together with poverty and overpopulation, urbanisation is one of the most significant processes affecting human societies in the twentieth century. Until recent times, urbanisation was almost universally seen as a direct indication of modernisation, development and economic growth. Throughout history, industrialisation and urbanisation have tended to occur together. But this simple monotonic relationship which has held for more than 6000 years, since the emergence of the very first cities, has changed quite fundamentally over the past four decades.

The world is currently experiencing an entirely new era of urbanisation. Today, it is the nations which make up the developing world which are experiencing the highest rates of urbanisation. How and why has this come about? In order to address this question, we need to look briefly at the whole history of world urbanisation.

The first settlements that can be referred to as urban date from the so-called **Urban Revolution**. This followed the **Neolithic Revolution**, which occurred in the Middle East some 10,000–8000 BC. It is important to understand the precise processes involved in this transition to urban living, for they are essentially the same forces that serve to shape the overall pattern of urbanisation in the contemporary global context. Accordingly, these processes are examined in detail at the beginning of Chapter 2.

For many centuries after the development of the first cities, the overall level of world urbanisation increased only very slowly and the urban areas which existed were small and effectively local in scope (Figure 1.1a and b). After the

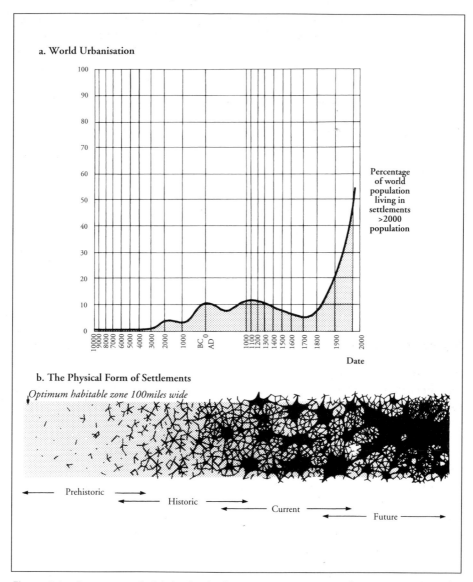

a. World Urbanisation

Percentage
of world
population
living in
settlements
>2000
population

Date

b. The Physical Form of Settlements

Optimum habitable zone 100miles wide

Prehistoric

Historic

Current

Future

Figure 1.1   A summary of global urbanisation, 10,000 BC to 2000 AD (source: Potter, 1985).

Urban Revolution, the subsequent history of urbanisation in the Middle East and Europe was complex and probably involved elements of both independent invention and spatial diffusion. However, it was with the rise of the great empires of the Greeks and the Romans, and to a lesser extent the Muslims, that urban life spread widely across Europe. The primary diffusion of the city that occurred under the Greeks was intensified under the Roman Empire in the first three centuries AD (Pounds, 1969). However, city life declined with the fall of the Roman Empire in the fifth century AD. In fact, it was not until the tenth and eleventh centuries that city development became important once

Table 1.1   Stages in the evolution of human societies

Stage no.	Stage	Approximate time period
1	Reciprocal societies	Up to 10,000 BC
2	Rank redistribution	10,000 to 3500 BC
3	Money-exchange systems	Up to 1400 AD
4	Mercantile societies	1492–1800
5	Capitalism: industrial and late	1850–2000+

Sources: after Johnston, 1980; Polanyi, 1968; Harvey, 1973

more, and it was the twelfth and thirteenth centuries that saw the rise of the medieval city, based on increasing local and long-distance trade. By the end of the Middle Ages, most of the major European cities of today were already in existence.

Understanding these and subsequent developments requires a brief overview of the development of human societies and the rise of the world economy. In his book *City and Society: An Outline for Urban Geography*, Ron Johnston (1980) recognised five broad epochs in the evolution of society: reciprocity, rank redistribution, money-exchange, mercantilism, and capitalism (both industrial and late), as summarised in Table 1.1.

The initial stage of the **Reciprocal Society**, the details of which will be fully explained in the next chapter, was synonymous with the first small societies that were of limited territorial extent. These settlement groupings were fully egalitarian and were based on consensus and democratic forms of decision-making. In them, exchange was based on reciprocal principles and no power or elite group existed. Such societies were pre-urban in all respects. Where productivity increased and a surplus product was first stored, the egalitarian structure of society broke down, being replaced by a rank-ordering of the members of society. At the same time, goods and labour were redistributed among members, so that the stage is referred to as that of '**Rank Redistribution**'. This is synonymous with the Urban Revolution, and the first emergence of military and religious power. Precisely why and how this should have occurred is considered in depth in Chapter 2.

Subsequently, trade was required in order to enhance economic growth, and was met by small-scale territorial expansion. Such a need was the precursor to the third societal change, identified as the **Money-Exchange** system, for trade necessitated the establishment of a common unit of exchange, or a monetary system. At this point, monetary rents could start to be charged by the owners of land and property for its use, rather than payments in kind. Johnston summarises how this led to a clear threefold division of society into an elite, a subject group and an intermediate set of administrators and military personnel.

However, in regard to global urbanisation, it was with the emergence of the fourth and fifth stages of the sequence, those of **mercantilism** and **capitalism** respectively, that major developments occurred. The principal hallmark of the mercantile society was the expansion of trade, and the outcome was the establishment of merchants whose job it was to articulate trade by buying and selling commodities, mainly over long distances. As the role of merchants demands that they buy goods first and then sell them, this required that they had access to capital before starting to trade. Surplus capital was now held by the owners of land, who could offer loans for merchants to establish themselves, so that a close interdependence could emerge between these two groups. Thus, some of the surplus previously resting entirely in the hands of the ruling elite was now shared with the merchant group. This early development of mercantile societies occurred in Europe around the fourteenth–fifteenth centuries and was associated with much more complex and interlinked settlement and economic patterns (see Figure 1.1a and b, page 5).

The most important development, however, came with **colonialism**, for continued mercantile growth required greater land resources. Historically, such growth occurred during the sixteenth to eighteenth centuries, first by means of trading expeditions, and then by distant colonialism, a trans-oceanic version of local colonial expansion, mainly involving areas of low socio-economic development. If an indigenous population already existed, colonial power could be enforced by administrative elites on tours of duty. In extreme cases, local populations were entirely wiped out, as in many Caribbean territories, where subsequently they were replaced by African slaves and indentured labourers (Lowenthal, 1960). In formerly unoccupied lands, a colony could be established, and under such conditions ports came to dominate the urban system, acting as **gateways** to the new colony. Frequently, a coastal–linear settlement pattern emerged, often characterised by the beginnings of strong urban concentration or **primacy** and **spatial polarisation**, a phenomenon which is further examined in Chapters 2 and 3. Thus, the settlement pattern of both the colony and the colonial power developed symbiotically from this point onward.

Whilst some commentators stress that the process of European colonisation of the traditional world by the Spanish, Portuguese, British, Germans, Belgians, French, Dutch and Italians represented the spread of economic growth and development, others argue that it represented a form of exploitation and expropriation, an observation that gives rise to ideas concerning modernisation and trickle-down processes on the one hand, and dependency and backwash on the other. These processes and associated philosophies of development are fully discussed in Chapters 2 and 3. Certainly the process involved the flow of extra profits or **surplus value**, often hailing from the production of a staple agricultural product such as sugar, first from rural areas to the colony's gateway primate city and thence to the major cities of the colonial power. These developments witnessed the increasing economic specialisation of countries and the international division of labour. An important conclusion is reached: that from this time onward, the economic evolution of the developed and developing worlds has been inextricably interlinked. Similarly, the rise of urbanisation can only be correctly interpreted if it is seen as a conjoint process involving the developed and developing worlds together. This argument of interdependent development and interdependent urbanisation (Brookfield, 1971; Roberts, 1978,

1995; Potter, 1985, 1992b) has important implications for both urban theory and practice, as will be demonstrated in this text.

The fifth historical societal stage identified by Johnston, **capitalism**, represents a complex development, but its principal outcome was that the scale of employment and production expanded dramatically with the rise of the factory system. Johnston (1980: 37) comments that 'although industrial capitalism is inextricably bound up with the factory system', its most important feature was, however, 'the alienation of labour power which enforces the working for wages'. Those with capital gained control over the means of production, whilst workers could only sell their labour. This first stage is known as **industrial capitalism**. As a subsequent stage, **late** or **monopoly capitalism** occurs when the potential market for goods becomes saturated. Under such circumstances, firms can only expand sales by capturing more of the existing market, so that successive mergers lead to a process of increasing industrial concentration.

It is the rise of industrial capitalism in the eighteenth century that brings us to the onset of the modern period of world urbanisation. Thus, the exponential growth of world urbanisation started only around 1800 (Figure 1.1a, page 5). At this point, only 3 per cent of world population was to be found living in urban settlements. The great growth in urbanisation came only with the Industrial Revolution in the late eighteenth and early nineteenth centuries in Europe. Davis (1965) has stressed the importance of the enormous growth in productivity that came with the use of inanimate sources of energy and machinery. Similarly, Sjoberg (1960) has emphasised the salience of the scientific revolution as the basis for the Industrial Revolution. He notes that the advent of industrialisation brought large-scale production, improvements in agricultural implements and farming techniques, improvements in food preservation techniques and better transport and communications. All of these technical developments in the initial period of the Industrial Revolution up to 1830 contributed to both increasing productivity in agriculture and the parallel development of industry. The steam engine was undoubtedly the key invention, leading to the rise of the factory system, whereupon mechanisation and mass production became established. Taken together, these fundamental changes allowed people to congregate in the evolving industrial city.

During the eighteenth and nineteenth centuries, development and change became even more interdependent at a global level, involving increasing links between industrial and non-industrial countries. This increasingly close relationship is expressed by what is referred to as **dependency theory** (Frank, 1967). The approach, which will be fully explored in Chapter 2, was largely derived in the context of Latin America and the Caribbean, and suggests that the development of Third World nations has been dictated by its integration into the capitalist mode of production, and that the more such nations have been incorporated into the global capitalist system, the more underdeveloped they have become (Frank, 1966, 1967; Beckford, 1972; Hettne, 1994), rather than the other way around.

The approach has clear implications for urban development in the form of what may be referred to as '**Dependent Urbanisation**' (Castells, 1977; Gilbert and Gugler, 1992), the basic components of which have been enumerated by Friedmann and Wulff (1976) when they state that developed countries established urban outposts in Third World countries for three closely related reasons. Firstly, to extract a surplus by way of primary products; secondly, to expand the market for goods developed under advanced monopoly capitalism; and thirdly, to ensure the continued stability of indigenous political systems that will most willingly support the capitalist system.

However one reacts to the central arguments of dependency theory, it is indisputable that the close interrelation existing between developed and developing countries from the sixteenth century onward has had a remarkable bearing on processes and associated patterns of world urbanisation. The epochs of mercantilism and industrial capitalism set the stage for the rapid urbanisation trends that were shortly to follow. Johnston (1980) observes that early industrial capitalism led to the accentuation of the primacy of gateway cities in the developed countries and served to bolster the divisions existing between levels of the settlement hierarchy. This argument is picked up in Chapter 2 in the form of Vance's mercantile model of settlement evolution and the closely associated plantopolis model (Rojas, 1989; Potter, 1995). A central conclusion is therefore that it was at this stage that the seeds of spatially unequal and polarised development were being laid in the countries of the developing world. Thus, it may be posited that without the intrusion of industrial capitalism and imperialism, some developing societies might still lack major cities today (Gilbert and Gugler, 1992).

Nineteenth- and twentieth-century patterns and processes of global urbanisation

The rapid urbanisation that occurred in Western Europe and North America during the late eighteenth and early nineteenth centuries was associated with a gradual process of industrialisation and economic change. During this period, a steady increase in the demand for labour occurred in the towns, whilst at the same time, technical developments in agriculture allowed for a declining rural population. Thus, a steady stream of migration occurred from the rural to urban areas. Health and sanitary conditions in the towns, however, were very poor indeed. As a result, there was little or no natural increase in urban populations. British cities were quite simply the death traps portrayed in the novels of Charles Dickens. For example, data from the Registrar General indicate that in 1840, life expectancy was only 24 years in central Manchester, 26 years in Liverpool and 28 years in London, against a national figure of 37 years.

The course of urbanisation in the developed world followed what can best be described as a smooth progression, whereby gradual demographic changes were matched by equally gentle changes in economic structure. This is well summarised by Kingsley Davis's (1965) classic concept of the 'Cycle of Urbanisation'. By examining the experience of a range of developed countries over the past 150 years, especially Britain, Davis argued that the progress of urbanisation is most accurately represented by a curve in the form of an attenuated

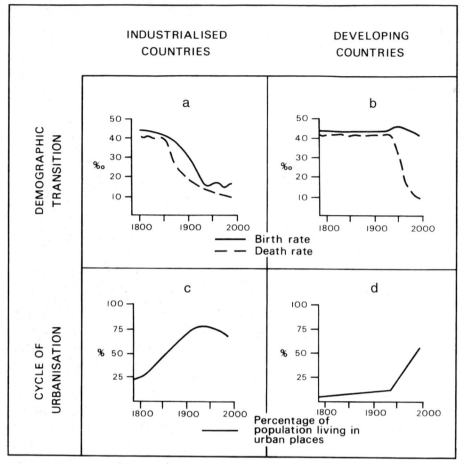

Figure 1.2   The cycle of urbanisation and the demographic transition for developed and developing countries (source: Potter, 1992b).

(or squashed) S-shape. Such a curve, shown in Figure 1.2c, is typical of many natural growth processes (see also Figure 1.3).

The generalised cycle envisaged for developed countries is shown in Figure 1.2c. Urbanisation proceeds gradually at first, but then there is a period of accelerated growth shown by the steep ascent of the curve. As the level of urbanisation climbs over 50 per cent, the curve begins to flatten out somewhat. After the 75 per cent level has been attained the curve levels out completely, or even declines with the onset of counter-urbanisation processes. Thus, urbanisation is seen as a finite process in that a saturation or ceiling point is assumed to exist.

The important point is that this relatively gradual process of urbanisation over a century and a half was directly paralleled by slow demographic change in developed countries (Figure 1.2a). Thus, the cycle of urbanisation and the demographic transition can be seen as running in harmony for developed

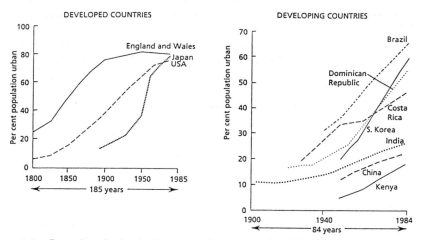

Figure 1.3 Examples of urbanisation curves for developed and developing countries (source: Potter, 1992b; after Davis, 1965).

countries over a considerable period of time. At the beginning of the nineteenth century, urbanisation was low and so was the rate of population growth. The period of most rapid urbanisation came in the early part of the nineteenth century, a time when the death rate dropped sharply and the population expanded quickly. As Great Britain moved into the late expanding stage of the demographic transition, urbanisation had slowed noticeably and, on entry into the low fluctuating stage, urbanisation had largely run its course. Thus, the process of urban development was a gradual one and Davis notes that it took some 79 years for England and Wales to move from a situation where 10 per cent of the population was urban, to one where 30 per cent were to be found living in towns and cities. In this connection Davis (1965) observed a general trend, for the later each country has industrialised, the faster it has urbanised. Hence, the 10 to 30 per cent change in urbanisation took 66 years in the case of the United States, 36 years for Japan, but only 26 years in the case of Australia.

The rapid rate of urbanisation that is currently affecting the Third World is a product of the post-1945 period. The majority of less developed countries are not following the cycle of urbanisation associated with the early stages of economic growth. They are urbanising more rapidly than the industrial nations did in their heydays. Davis (1965) calculated for the underdeveloped countries for which data were available in the 1940s and 1950s, that the average annual gain in urban population was 4.5 per cent. In comparison, for nine European countries during their periods of peak urbanisation, the annual average urban gain had been 2.1 per cent. Thus, as shown in Figure 1.2d, most developing countries displayed a steep, almost vertical urbanisation curve after the Second World War.

Just as many developing countries have telescoped the cycle of urbanisation (Figure 1.2d), so they have frequently foreshortened the demographic transition (Figure 1.2b). A major feature of this group of countries has been a very

Plate 1.1   The hustle and bustle of a street market in Old Delhi, India (photo: Rob Potter).

rapid growth of total population. In the period 1970–75, the population of
the world grew at an overall rate of 1.9 per cent per annum. However, this rate
comprised a growth rate of 2.4 per cent per annum in the less developed realm,
and one of 0.9 per cent per annum in the developed world.

There is, however, another very important distinguishing feature of urbanisation in the developing world and this is the huge *absolute* numbers that are
involved. This is mainly the outcome of a sharp decline in the death rate due
to better medical and health care facilities in the post-war period, whilst birth
rates have remained at traditionally high levels (Figure 1.2). Hence, there is
massive growth of total population, both rural and urban, in many Third
World countries (Plate 1.1).

Thus, even if there were no relative shift in the proportion of developing
world countries' populations living in urban areas, that is urbanisation *per se*,
rapid urban growth would still occur due to the very high rates of natural
increase of population which are involved. As health and social welfare standards are generally so much better in the cities than in the rural areas, Third
World cities exemplify *par excellence* the combination of pre-industrial fertility
with post-industrial mortality. Contemporary cities in the developing world
exhibit some of the highest rates of natural increase ever found in cities (Dwyer,
1975; Davis, 1965). For example, in Caracas, Venezuela, from 1960 to 1966,
52 per cent of the population increase was accounted for by in-migrants, whilst
an identical level was recorded for Bombay, India for the period 1951–61.
MacGregor and Valverde (1975) noted that 31 per cent of the total population
of Monterey and 48 per cent of Tijuana were in-migrants in the early 1970s.

In fact, very often, natural increase and in-migration contribute in broadly equal proportions to the total growth of urban populations. Thus, in the Philippines in the 1970s, the in-migration rate to cities was 1.8 per cent per year, out of a total population growth per annum of 3.9 per cent. For Brazil during the same decade, in-migration accounted for 2.2 per cent per annum out of a total urban growth rate of 4.4 per cent per year (United Nations, 1988; Devas and Rakodi, 1993).

But relative rural–urban shifts in population are occurring too, so that Third World cities are growing at very rapid rates. Often, the percentage annual growth of urban population is running in excess of 6 per cent per annum, and frequently over 10 per cent. At the latter rate, the absolute number of urban dwellers will double every seven years, and at the former, every 10 years. During the decade 1960–70, for example, the percentage annual growth of urban population for Malawi was 10.1, for Tanzania 8.6, Nigeria 6.0, Malaysia 5.9, Venezuela 5.6 and Colombia 5.0.

Why has rapid urbanisation become so characteristic of contemporary developing countries? Customarily this is explained by a combination of 'push' and 'pull' factors. With regard to pull factors, the enhanced health care facilities provided in cities in the developing world has already been stressed (see Phillips, 1990). In addition, after the Second World War, many developing countries were encouraged, and were themselves keen, to seek economic growth and prosperity by means of industrialisation, especially programmes based on import substitution industrialisation, for example the production of soft drinks and the like. Historically, as noted earlier, markets and infrastructure have normally been centred on the primate gateway cities of developing societies. Thus, post-war development planning and the legacies of colonial and capitalist penetration have led to the increasing concentration of industrial and commercial job opportunities in the major urban areas. For example, in the case of Nigeria, some 76 per cent of all manufacturing employment is to be found in the narrow coastal zone forming the south of the country, running from Lagos to Port Harcourt, and Calabar (Filani, 1981; Potter, 1992b).

Roberts (1978, 1995) echoed traditional dependency theory when he argued that it is the form of industrialisation that has occurred, and not inertia or traditionalism, that is holding back development in much of the Third World. Others, such as Lipton (1977, 1982), although accused of over-simplification, have maintained that Third World poverty is largely the product of **urban bias** in all aspects of development.

The industries established in developing countries in the post-war period were often of the 'final touch' variety, involving the final stages of the production sequence, and hence provided relatively few jobs. Later, the attraction of branch plants of overseas multinationals was attempted, by means of offering tax breaks and the suspension of planning and other regulations. What is undeniable, however, is that the jobs that are available offer relatively high wages in comparison with rural incomes. Further, as elsewhere, average income levels tend to rise progressively with increasing city size (Hoch, 1972). Hence, cities in the developing world have come to offer the *hope* of employment, high wages, better opportunities and an improved environment, if not immediately, then in the future.

In addition, there is also the existence of the **informal sector** of the economy – small-scale and individually led – which also offers the chance of establishing a toehold in the urban economy (Santos, 1979; McGee, 1979a; Portes *et al.*, 1991). Further, many would maintain that in the past, inappropriate curricula in schools, colleges and universities have also served to orient young people towards white-collar urban-based occupations (Mabogunje, 1980). Hence, the Third World city is essentially a set of perceived opportunities, so that 'perhaps they do no more than promise the *hope* for work and settlement; but to obtain even crumbs one must be near the table' (Jones and Eyles, 1977: 210). Some occasionally observe that the overt poverty, crowding, unemployment and sporadic squalor of Third World cities should act as deterrents to migrants but, of course, this argument ignores the massive unemployment and poverty which are to be found in the rural areas of so many developing nations (Gilbert and Gugler, 1992; Potter, 1985, 1992b). In fact, as will be argued in Chapter 3, cities are based on the very existence of such inequalities.

Statistics reveal the precise degree to which urbanisation in the developing world has been outrunning the rate at which it has been experiencing industrialisation. This disparity is frequently referred to as '**urban inflation**' or '**hyper-urbanisation**'. Statistics included in Bairoch's (1975) survey of the economic development of the Third World in the twentieth century exemplify this trend. In 1970, in the non-communist less developed countries taken as a whole, 21 per cent of the population was urban, whilst some 10 per cent of the active population was engaged in manufacturing, so that urbanisation exceeded manufacturing employment by 110 per cent. By comparison, for the countries of Europe in 1930, 32 per cent of the population had been urban and 22 per cent was employed in manufacturing, giving an urbanisation surfeit of only 45 per cent.

Despite the strong efforts made by developing nations to industrialise, manufacturing still only accounts for a relatively small part of the economies of such nations. In 1960, industry accounted for some 15.6 per cent of the total gross domestic product of developing countries taken as a whole. By 1983, the level of industrialisation had increased, but only marginally, to 17.5 per cent. At this latter date, industry accounted for approximately 25 per cent of production. Since 1983, levels of industrialisation have increased. In 1994, industry accounted for 36 per cent of gross domestic product for developing countries taken as a whole (World Bank, 1996). But even more saliently, the development of industry has been highly uneven between the different parts of the developing world. Latin America and the Caribbean, together with East and Southeast Asia, represent the most industrially developed areas of the Third World. Turning to individual countries, Korea, Hong Kong, Mexico, Brazil and Singapore together accounted for just under 60 per cent of all developing country manufacturing exports in 1986 (Chandra, 1992). These data serve to pinpoint the current degree to which the growth of towns and cities in less developed nations is running ahead of industrialisation, the provision of jobs, infrastructure, welfare services and adequate housing.

Collectively, these conditions represent the push and pull factors stimulating urban growth and urbanisation in developing societies. Thus, rates of urbanisation as opposed to levels of urbanisation in the world are increasingly becoming inversely related to levels of economic development and wealth, altering in dramatic fashion a relationship which has previously held for some five and a half millennia. It is the developing countries that account for the extremely rapid current rate of world urbanisation, whilst the developed nations are exhibiting declining rates, and some even declining levels of urbanisation. Thus, Dwyer, writing in 1975, observed that in 'all probability we have reached the end of an era of association of urbanisation with Western style industrialisation and socio-economic characteristics' (Dwyer, 1975: 13). This is certainly the case, and provides the context in which we must consider the nature and role of the city in the developing world.

The city in the developing world: contemporary context and predictions to 2050

The statistics included in Table 1.2 show that in the early 1920s, there were 24 cities in the world with more than one million inhabitants. Sixty years later, by the early 1980s, the number of so-called '**million cities**' had increased to 198. But far more significantly, during each decade between these two dates, the average latitude of million cities had moved steadily towards the equator (Table 1.2). The mean location of million cities had shifted from 44 degrees 30 minutes in the 1920s to 34 degrees 7 minutes by the early 1980s, showing that million cities are increasingly associated with the tropical and sub-tropical realms. The growth of very large or 'mega cities' has attracted much attention over the last 10 years or so (see Oberai, 1993; Gilbert, 1996).

Further, by 2000, it is believed that just under half of all urban dwellers in developing countries will live in cities of one million or more (Harris, 1989). In 1950, there were 31 cities of a million or more inhabitants in developing countries, but by 1985 there were 146. It is projected that by 2025 there will be a staggering 486 cities of a million or more in less developed countries (Harris, 1989). All of these facts serve to illustrate the degree to which the

Table 1.2   The world distribution of cities with more than one million inhabitants, 1920s–1980s

Date	Number of 'million' cities	Mean latitude north or south of equator	Mean population (millions)	Percentage of world population living in 'million' cities
Early 1920s	24	44°30'	2.14	2.86
Early 1940s	41	39°20'	2.23	4.00
Early 1960s	113	35°44'	2.39	8.71
Early 1980s	198	34°07'	2.58	11.36

Source: Potter, 1992b

Table 1.3   The largest cities in the world, 1950 and 2000

	1950			2000	
Rank	City	Population (millions)	Rank	City	Population (millions)
1	New York	12.3	1	Mexico City	31.0
2	London	10.4	2	São Paulo	25.8
3	Rhine–Ruhr	6.9	3	Shanghai	23.7
4	Tokyo	6.7	4	Tokyo	23.7
5	Shanghai	5.8	5	New York	22.4
6	Paris	5.5	6	Beijing	20.9
7	Buenos Aires	5.3	7	Rio de Janeiro	19.0
8	Chicago	4.9	8	Bombay	16.8
9	Moscow	4.8	9	Calcutta	16.4
10	Calcutta	4.6	10	Jakarta	15.7
11	Los Angeles	4.0	11	Los Angeles	13.9
12	Osaka	3.8	12	Seoul	13.7
13	Milan	3.6	13	Cairo	12.9
14	Bombay	3.0	14	Madras	12.7
15	Mexico City	3.0	15	Buenos Aires	12.1

Source: United Nations, 1989a

world's largest urban places are becoming a feature of developing, rather than only developed, countries.

The trend towards developing world urbanisation is also exemplified by the league table of the largest urban places in the world in 1950 and 2000 (Table 1.3). In 1950, the three largest cities in the world – New York, London and the Rhine–Ruhr conurbation – were all in the developed world. By 2000, by contrast, it is anticipated that three developing world cities, Mexico City, São Paulo and Shanghai, all with over 23 million inhabitants, will represent the world's largest cities. At this latter date, of the 15 largest cities in the world, 12 can be described as being in the Third World. This compares with only five in 1950. Statistics contained in the United Nations volume *Prospects for World Urbanization 1988* (United Nations, 1989a) indicate that between 1970 and 1985, Mexico City was growing at an average annual rate of 4.30 per cent, and although this is likely to reduce somewhat, it will remain at 2.56 per cent per annum between 1985 and 2000. São Paulo in Brazil was growing at 4.38 per cent per annum between 1970 and 1985, and will continue at 2.79 per cent between 1985 and 2000. Meanwhile, in Korea, Seoul was growing at 4.27 per cent per annum from 1970 to 1985, and is projected to continue at a rate of 1.69 per annum between 1985 and the end of the century.

The world's fastest growing cities during 1985–2000 are shown in Figure 1.4. It is noticeable that these fast growing cities are all located to the south of the line which is customarily drawn between the rich 'North' and the poor

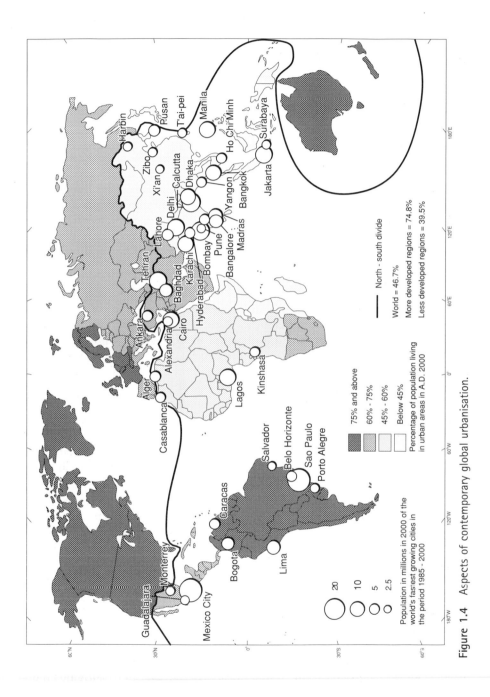

Figure 1.4 Aspects of contemporary global urbanisation.

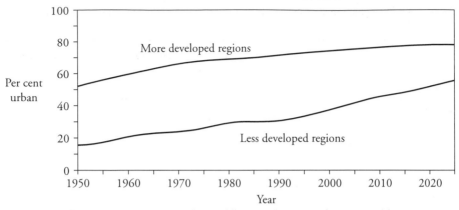

Figure 1.5 The proportion of population residing in urban areas in more and less developed regions, 1950–2025 (adapted from United Nations, 1989a).

'South'. Cities such as Karachi in Pakistan, as well as Lagos, Kinshasa and Addis Ababa in Africa, grew by more than 50 per cent between 1985 and 2000.

The overall increase which has occurred in the level of urbanisation between 1950 and 2025 is summarised for the more developed and less developed regions of the world in Figure 1.5. A disaggregation of these data for major continental divisions of the developing world is detailed in Table 1.4. The clearest trend is shown in the graph depicted in Figure 1.5. Thus, whilst the urbanisation curve has flattened out for the developed world since 1965, it has increased sharply in the case of the developing world. By 2000, 74.8 per cent of the population of more developed countries will be urban, whilst this will be true of 39.5 per cent of those living in less developed regions. By 2025, these proportions will have risen to 79.0 per cent for the developed world and 56.7 per cent for the developing world.

Some very important outcomes of this rapid process of urbanisation in the developing areas of the world are witnessed by the data shown in Table 1.4. By 2025, it is envisaged that just under 60 per cent of all Africans will be living in urban settlements, as will just over half of all those living in Asia. In the same year, nearly 85 per cent of all Latin Americans will be living in towns and cities. This is all a far cry from the situation in 1920, when United Nations data indicate that less than 10 per cent of Africans and Asians, and only 22 per cent of Latin Americans, were urban denizens (Table 1.4).

These relative percentage figures, however, tell only part of the story, for the real magnitude of the process of urbanisation is only revealed if we look at the absolute numbers involved. This is shown in Figure 1.6 (page 20) and Table 1.5. Up to 1970, the absolute number of city dwellers was larger in the more developed regions than in the developing regions. However, from this year onward, the number of city dwellers has risen dramatically in the developing world. During 1950–2025, the number of urban dwellers in developing countries will have increased 14-fold, from 300 million to a staggering 4 billion. In 2000, the statistics indicate the existence of two city dwellers in the developing

Table 1.4 Percentage of population living in urban areas by major continental region, 1920–2025

Region	Percentage of total population living in urban places			
	1920	1970	2000	2025
World:	19	37.2	46.7	60.5
More developed regions	40	66.6	74.8	79.0
Less developed regions	10	25.5	39.5	56.9
Africa:	7	22.9	41.3	57.8
East Africa	–	10.3	30.1	48.0
Middle Africa	–	24.8	47.6	64.7
Northern Africa	–	36.0	49.9	65.3
Southern Africa	–	44.1	61.7	74.2
Western Africa	–	19.6	40.7	58.9
Latin America:	22	57.3	77.2	84.8
Caribbean	–	45.7	65.5	75.5
Central America	–	54.0	71.1	80.5
South America	–	60.0	81.0	87.5
Asia:	9	23.9	35.0	53.0
Eastern Asia	–	26.9	32.6	49.0
Southeast Asia	–	20.2	35.5	54.3
Southern Asia	–	19.5	33.8	52.6
Western Asia	–	43.2	63.9	76.3

Sources: United Nations, 1988, 1989a; UNCHS, 1996

Table 1.5 Numbers of people (in millions) living in urban and rural settlements in 2000 and 2025 by major continental region

Region	2000			2025		
	Urban population	Rural population	Total population	Urban population	Rural population	Total population
World:	2,916	3,334	6,251	5,118	3,347	8,466
More developed	944	317	1,262	1,068	283	1,352
Less developed	1,971	3,016	4,988	4,050	3,063	7,114
Africa	360	511	872	913	667	1,580
Latin America	416	122	539	644	115	760
Asia	1,292	2,405	3,697	2,589	2,300	4,889
North America	221	73	294	259	73	332
Europe (excl. former USSR)	386	121	508	421	90	512
Former USSR	217	90	307	260	91	351
Oceania	21	9	30	29	10	39

Sources: United Nations; 1988, UNCHS, 1996

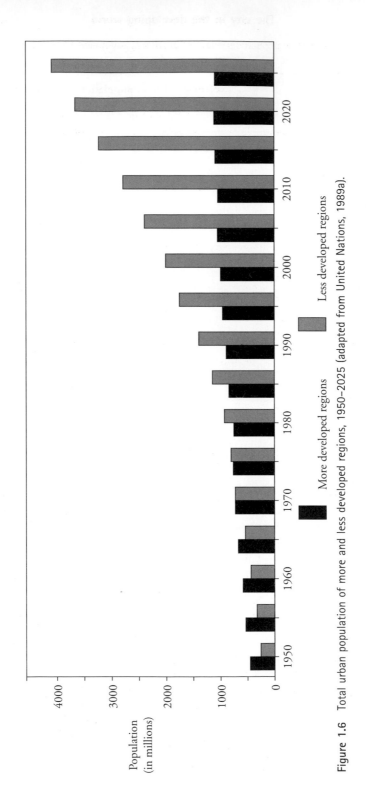

Figure 1.6   Total urban population of more and less developed regions, 1950–2025 (adapted from United Nations, 1989a).

Table 1.6 Average annual percentage rate of growth of urban and total populations by major continental division, 1995–2000 and 2020–2025

Region	1995–2000		2020–2025	
	Urban population	Total population	Urban population	Total population
World:	2.6	1.6	1.9	1.0
More developed	0.7	0.4	0.3	0.2
Less developed	3.6	1.9	2.4	1.1
Africa:	4.8	3.0	3.0	1.9
East Africa	6.1	3.3	3.7	2.1
Middle Africa	4.8	3.0	3.1	2.1
Northern Africa	3.6	2.3	2.2	1.3
Southern Africa	3.2	2.2	1.9	1.3
Western Africa	5.3	3.3	3.2	2.0
Latin America:	2.4	1.8	1.4	1.1
Caribbean	2.2	1.3	1.2	0.7
Central America	2.7	2.0	1.7	1.3
South America	2.3	1.8	1.3	1.1
Asia:	3.4	1.7	2.3	0.9
Eastern Asia	2.3	1.1	1.9	0.4
Southeast Asia	3.7	1.7	1.0	0.9
Southern Asia	4.2	2.2	2.6	1.1
Western Asia	3.5	2.6	2.3	1.7

Sources: United Nations, 1988; UNCHS, 1996

world for every one in the developed world. And by 2025, the data and graph show that the ratio will have increased to almost four city dwellers in the developing world for every one to be found in the developed world. Table 1.5 (page 19) also shows that it is predicted that by 2025, there will be an astounding 2589 million urban inhabitants in Asia alone, more than double the number that will be found in the developed world taken as a whole. By the same date, Africa will have some 913 million urban residents, far more than those to be found in Europe (421 million) and North America (259 million).

By the end of the first quarter of the twenty-first century, therefore, urban living will have become firmly associated with the poorer countries of the globe. Whilst the rate of urban population growth has been running at 3.6 per cent per annum between 1995 and 2000 in the developing world, during the same period the rate was 0.7 per cent per annum in the more developed world (Table 1.6). These growth rates will decrease to 2.4 per cent per annum for the developing world and 0.3 per cent for the developed world respectively between 2020 and 2025 (Table 1.6). The highest predicted urban growth rates between 2020 and 2025 pertain to eastern Africa, where it is predicted that the rate will be 3.7 per cent per annum. The annual rate of urban population growth will also remain above the 3 per cent level in both middle Africa and western Africa during this period (Table 1.6).

What of the physical form that such cities will take on the ground? A good starting point in examining this question is a book published in the early 1970s by two Greek urban scholars, Doxiadis and Papaiouannou (1974), in which they forecast the development of what they called 'Ecumenopolis', or a single functional world city, by the middle of the twenty-first century. The idea of Ecumenopolis began with the coalescence of major urban areas in developed countries into massive polynuclear urban forms. On the northeastern seaboard of the United States, Gottmann (1957) recognised in the 1950s what he referred to as 'Megalopolis'. This comprised a whole chain of metropolitan areas stretching the 600 miles between Boston and Washington, and which taken together housed 30 million people.

Subsequently, Gottmann (1978) recognised six such megalopolitan systems around the world, each with a population in excess of 25 million. These included two in the developing world, in Brazil based on Rio de Janeiro/São Paulo, and in China focusing on Shanghai. If a 10 million population threshold is taken, many other megalopolitan systems can be recognised, including ones centred on Buenos Aires, Calcutta, Bombay, Cairo and Mexico City in the Third World.

Doxiadis and Papaiouannou's suggestion is that these complexes will eventually become linked in a chain-like form, giving a series of interrelated megalopolitan systems (Doxiadis, 1967). Thereby, they envisaged that high density urban lineaments will eventually connect the urban cores within Latin America, Africa and Asia. The term 'Ecumenopolis' is used to indicate a functionally integrated urban whole and is clearly not meant to imply complete physical coverage of the world's land surface. In order to produce their forecast, Doxiadis and Papaiouannou assumed world populations of 6430 million by 2000 and 9600 million by 2050, of which 71.5 per cent would be residing in cities. They then explored the habitability of different areas of the globe in 2100, according to climate, altitude and water supply, and thereby ascertained areas for possible future urban development. These zones were then used to define a theoretical configuration of global urban centres and growth axes. The work subsequently projected the likely settlement pattern, assuming world populations of 20,000 million by 2100 and 50,000 million by 2200.

The authors subtitled their book 'the *inevitable* city of the future'. Although some may remain unconvinced of the unavoidability of Ecumenopolis, its prediction is an interesting physical corollary of the statistical projections on global urbanisation presented in this chapter. Further, the forecast of Ecumenopolis once again makes us aware of the highly interrelated and interdependent nature of the urban process in both the developed and developing worlds. However, this should not be taken as implying that the cities of the poor and rich parts of the world will come to look identical – far from it. This theme is pursued at the level of individual cities in Chapter 6, having been considered previously at the national/regional scales in Chapter 3. For example, a number of urban scholars have argued that although complex, functionally integrated zones can increasingly be recognised in Asia, these are characterised by the development of a growing tendency for non-agricultural and agricultural activities to be found alongside one another surrounding the large urban cores of many Asian countries, at both the spatial and household levels. In a deliberate attempt to distance the concept from Eurocentric western connotations, McGee (1989) coined the term 'Kotadesasi', by juxtaposing the Indonesian words for town and country, to describe such emerging urban forms (see also Chu-Sheng Lin, 1994). More recently elsewhere, McGee has referred to *desakota* regions (see McGee, 1991b, 1995; Firman, 1996) and to the growth of Extended Metropolitan Regions (EMRs) (McGee and Greenberg, 1992) in Pacific Asia. From a similar perspective, Baker (1995) and Benthall and Corbridge (1996) have explored the nature of rural–urban interactions in Tanzania and India.

Different patterns are, however, likely to appear in other parts of the developing world, especially given the demise of state socialism following its collapse in the former USSR and in eastern Europe since 1989. It may be posited that a possible outcome of these transformations is the intensification of capitalist development and the further gravitation of capital to urban cores (see Potter and Unwin, 1995). In these circumstances we might expect a very different outcome from that just described for parts of Asia. In particular, a growing polarity and distinction between the urban and rural sectors of the economy might well be the outcome, with this disjunction being increasingly predicated on strong and growing rural to urban flows (Potter and Unwin, 1989, 1995). The fact that many leading aid agencies, but in particular the World Bank, have recently stressed afresh what they see as the need for urban-based increases in productivity to fuel the development process, makes this scenario increasingly likely wherever the capitalist path is followed in earnest (see Gould, 1992; Harris, 1989, 1992). This theme is picked up in several parts of the text and forms part of the conclusions presented in Chapter 10.

Can we generalise about urban conditions in the developing world?

This brings us to the final section of this chapter. Given the interdependence of the developed and developing worlds, and the manifest heterogeneity of different parts of the globe and of the individual nations which make up these broad divisions, can, and indeed should, we try to make generalised statements about the developing world as a whole? A useful offshoot of this discussion is a consideration of the nature and definition of the developing world itself, something that so far we have not attempted.

There is an argument that the developing world is so diverse that it is at best highly misleading to endeavour to make generalisations concerning conditions within it, and at worst entirely fruitless (see Plate 1.2). Certainly, cultural diversity alone is likely to negate any efforts to derive neat universal solutions to social, economic and political problems spanning the entire Third World. Thus, Gilbert and Gugler (1992: 220, 221) note how 'too often policies useful in one country at one specific time are turned into panacea for all countries at

Plate 1.2   Modern transport intrastructure and buildings in Hong Kong (photo: Rob Potter).

all times', and that 'instant solutions taken from the latest vogue generalisation have wrought havoc' in the field of planning and development. Examples in relation to urban planning and development that will be looked at in some detail in this book include the idea of spreading development by means of so-called 'growth poles' and 'growth centres', a notion which came into vogue in the 1970s, along with ideas of classic hierarchical settlement systems and modernisation theory as a standard template for all nations (see Chapters 2 and 3). Thus, it should be stressed that in considering developing countries as a whole, we do not mean to imply that universally applicable concepts and policy responses can be identified.

Further, the essential heterogeneity of the countries which make up what is commonly referred to as the 'developing world', the 'Third World', the 'less developed world' or whatever, has to be recognised right at the outset. The term 'Third World' was coined during the Cold War period of the 1950s, and despite its common employment, heated debate still surrounds its use (see O'Connor, 1976; Auty, 1979; Wolf-Phillips, 1979; Mountjoy, 1980; Ward, 1980; Drakakis-Smith, 1993; Milner-Smith and Potter, 1995). The term originally had a purely political meaning, serving to denote those countries which were non-aligned at the end of the Second World War. These were the newly independent nations escaping from colonialism, which were not committed either to the western free-market bloc or to the eastern socialist centrally planned group of countries at that juncture. Today, however, the term seems increasingly to be used to denote the poorer countries of the world (Milner-Smith and Potter, 1995). With the demise of much of the 'Second World' since the break-up of the Soviet Union, many see the terms First and Third World as being rendered equally obsolete (see Drakakis-Smith, 1992).

The present text generally uses the expressions 'developing countries' and 'developing world' as shorthand for the set of countries which are relatively poor. Most of these countries are to be found to the south of a line drawn around the globe which serves to include Latin America, the Caribbean, Africa and most of Asia. This line is shown in Figure 1.4 (page 17), with the possible exception of richer states such as Israel, Hong Kong and Singapore (Gilbert and Gugler, 1992: 6). Most, but not all, of these countries make up the tropical and sub-tropical world, and they disproportionately count amongst them former colonies. More saliently, they represent the 75 per cent of the world's population that accounts for only 15 per cent of its income, 14 per cent of world industry, and 20 per cent of global energy consumption (Potter, 1994: 113).

The central argument advanced in this book is that it is fitting to consider the developing world as a whole, just as long as we recognise that the problems faced by such countries are different in *scale*, rather than in *kind*, from those that are faced by the richer nations of the world. Hence, problems of regional imbalance and inequality, social polarisation, urban concentration, unemployment, poor housing and access to services and structural poverty occur in all societies. But they affect the poor in the poorer countries more than the relatively well-off and the poor in the rich world. Such an observation also relates to the important argument that poverty is a *relative* as well as an *absolute* phenomenon, and that from the point of view of social policy, it is the occurrence of inequalities that is more important than poverty *per se*. This argument will be highlighted at several points in this volume, but especially in relation to the provision of housing, other basic needs and employment in Chapters 5–8.

Some have put forward the argument that the magnitude of the problems faced by developing countries means that planning and development initiatives have to be premised on different foundations. Thus, Taylor and Williams (1982) note that planning in developing societies needs to stress social planning rather than physical (land use) planning, and is more likely to be based on local initiatives, rather than comprehensive master plans. Further, planning in developing areas has to be predicated on the fact that people are having to rely on **self-help** much more than in developed countries, and that the employment and housing markets are as a result, far more **informal** in these contexts (Koningsburger, 1983; Potter, 1985). The fact that these nations face the same problems as other more developed ones, but in a far more pressing form, and that the severity of problems also varies sharply between nations within the so-called developing world, means that we can, and should, seek to generalise about the experiences and prospects of developing countries. In short, the problems of poverty and inequality are so pressing and exhibit so many commonalities that consolidated and generalised approaches are required.

However, as Auty (1979) and O'Connor (1976) have reminded us, there is really a wide variety of developing countries which are best regarded as making up a continuum. Thus, there are great differences even within the developing world in terms of the size of countries, their resource bases and their demographic circumstances. Auty (1979) distinguishes between categories such as the oil-rich OPEC countries, the Newly Industrialised Countries (NICs), and the small impoverished island states of the Pacific and the Caribbean, in arguing the case for a seven-fold division of the developing world. O'Connor has argued the case for a six-fold typology, whilst Wolf-Phillips (1979) recognised four worlds. The emergence of the post-communist states since 1990 has added yet another complication to these already complex typologies.

Summary

Urbanisation is occurring much more rapidly in today's poorer countries than it did in the more developed nations during the heyday of the Industrial Revolution. Clearly, urbanisation as a global process can no longer be seen as a direct correlate of development and modernisation. The historical account of urbanisation included in this chapter stresses, however, that the evolution of the developed and developing worlds has to be seen as inextricably linked. This interdependence is also shown physically with the prediction of 'Ecumenopolis', or a functionally integrated world urban system, by the end of the first quarter of the twenty-first century. The prediction of Ecumenopolis does not mean, however, that everywhere will become the same – far from it, as subsequent chapters in this book will amply demonstrate. But the fact that the poorer countries of the world which together make up the developing world are facing such pressing problems of structural poverty and inequality, means that considering them as a whole is not only possible, it is a pressing priority of immense significance to the future habitability of the planet.

Chapter 2

Third World urbanisation and development: theoretical perspectives

Introduction

A pivotal argument presented in this chapter is that in order to understand the process of urbanisation, it is necessary to ponder the process of development itself. In a sense this argument was exemplified in the account of global urbanisation which was presented in Chapter 1. In the present chapter, however, the principal aim is to outline the theoretical foundations and implications of this argument. For example, the significance of so-called '**top-down**' planning and development can only be appreciated when it is recognised that the term refers to an ideology of change which is based on the belief that development best occurs at the top of the settlement hierarchy, and only then filters downwards. In contrast, the more recently emphasised, although considerably more diverse, philosophies of '**bottom-up**' planning and development connote situations where it is argued that change should focus initially on the lower echelons of the settlement system, and only subsequently be transmitted up the settlement hierarchy. This vitally important, although quite straightforward, distinction exemplifies the theme of the present chapter, namely, that the processes of urbanisation and development go hand in hand and need to be considered together.

The origins of urban living and the role of a social surplus

Preamble

Humans have lived in settlement clusters of sufficient size to occasion use of the label 'urban' for at least 5500 years. However, as was briefly indicated in Chapter 1, once established, the spread of urban settlements and predominantly urban modes of life was a slow process. As Figure 1.1 (page 5) has illustrated, by 1800, nearly five and a half millennia after the development of the first true cities, only an estimated 3 per cent of the total world population was to be found residing in towns and cities.

Given the length of the transformation, why should we be concerned with it? The answer to this important question is that we can only hope to develop a realistic appreciation of today's global and regional processes of urbanisation if we examine the factors that are believed to be behind the very first impulse towards urban living. This was well summarised by Bird (1977: 27) when he commented that the 'study of city origins throws a great searchlight on the march forward of human society'. Despite the involvement of archaeologists, anthropologists and historians in the study of urban development, a surprising amount of writing in the field has been disappointingly ahistorical. Thus, Friedmann and Wolff (1976: 10) warned of what they regarded as the 'facile generalisations of those social scientists who are inclined to think that the start of urbanisation in Third World countries coincided . . . with the beginnings of their own interest in the study of this process'. It is important that this warning should be heeded. Urbanisation is by no means a process which is new to the countries which make up the contemporary Third World. Indeed, the first regions of urban development were all located in what today are considered to be part of the developing world.

What were the circumstances whereby humans first started to live in large, dense and permanent agglomerations associated with fundamental economic, demographic, social and behavioural transformations? As noted in Chapter 1, such changes were so far-reaching that they are referred to as the second or '**Urban Revolution**' which followed the first or **Neolithic Revolution** (Childe, 1950, 1951; Adams, 1960; Sjoberg, 1960). This is the same as the change from reciprocity to rank redistribution within society, involving the development, storage and redistribution of a social surplus, as outlined in Chapter 1. It is understanding the nature of this so-called **social surplus product** which is vital to our understanding of the city in the contemporary developed and developing worlds. As Davis (1973: 9) commented, 'the theory of how cities began is beginning to be integrated with the theory of how cities operate in modern society'.

When and where did the first cities develop?

The current view is that there were at least seven regions of **primary urban generation**, that is, areas of apparently independent urban development (Wheatley, 1971). These are all in the present-day Third World: (1) Mesopotamia, (2) Egypt, (3) the Indus Valley, (4) the North China Plain, (5) Meso-America, (6) the Central Andes, and (7) southwest Nigeria. Together these areas may be regarded as forming the so-called **pre-industrial civilisations**.

The standard account regards the first true cities as having developed in the 'cradles of civilisation', or the 'fertile crescent' made up by the Tigris, Euphrates and Nile riverine valleys of the Middle East. According to Sjoberg (1960), the rise of urban civilisation dates from 3500–3000 BC in Mesopotamia, among the earliest cities being Eridu, Ur, Lagash, Larsa, Kish, Jemdet Nasr and Uruk. With regard to the Nile Valley, the main period of urban development dated from 3100 BC, and was linked with the dynasties of the Pharaohs, centred on such cities as Memphis and Thebes. These were largely mortuary cities associated with the pyramids and temples, and some argue that such settlements were quite small, perhaps reflecting the practice of changing the site of the capital with the ascendancy of each new Pharaoh.

Initially, archaeologists such as Childe (1951) suggested that the so-called urban revolution occurred only in the ancient Near East and that later developments in other regions represented a form of spatial diffusion from this original area. But the more recent position is that whilst specific items of technology may well have been diffused, the actual development of urban life arose independently in these other regions.

Thus, cities in the Indus Valley appear to have developed around 2500 BC, two well-documented examples being Mohenjo-Daro and Harappa in modern-day Pakistan. Both settlements show evidence of irrigation and were well planned on a regular rectangular grid basis. However, urban life in this area appears to have atrophied and ceased altogether after 1500 BC.

In China, urban development is thought to have started around 1500 BC on the alluvial plains of the Yellow River, with Anyang and Chengchow representing two important urban places. In the case of meso-America, ceremonial complexes such as that at Teotihuacan in Mexico seem to date from around 1000 BC. In the central Andes, settlements based on maize and shifting cultivation are dated from 500 BC. However, some archaeologists doubt that these were true cities and argue that they were merely ceremonial foci for low-density rural populations in the vicinity. Finally, there is evidence that the Yoruba territories of present-day Nigeria constitute a somewhat later region of primary urban generation. It is suggested that ceremonial centres may have appeared as early as the end of the first millennium AD, the initial development being at present-day Ife. It is believed that other subsequent areas of development such as Crete, southeast Asia and Etruria were all secondary, having been derived from these seven primary areas of urban generation (Wheatley, 1971).

The processes involved in the rise of the first cities

Even this brief résumé of pre-industrial urban development gives clear pointers to the hypotheses that may be advanced to explain urban genesis. Widening the discussion somewhat, Carter (1977, 1983) argues that four main explanations for the initial emergence of towns and cities may be identified. Firstly, there are **hydraulic** or **environmental–ecological theses** suggesting that cities occurred due to the presence of a favourable physical environment which allowed the extraction of an agricultural surplus. **Economic theories**, on the other hand, imply that the city was a product of the articulation of long-distance trade and regional market functions. Thirdly, it may be posited that towns grew for **military purposes** at defensive strong points. Lastly, **religious theories** envisage that urban development occurred about the foci afforded by shrines and temples.

But before turning to consider these various explanations in greater detail, the precursors to the urban revolution need briefly to be considered. During the Palaeolithic or Old Stone Age, a period generally equivalent to the Pleistocene geological epoch, human groups relied entirely on hunting, fishing and gathering. Then, during 10,000–8000 BC, the **Neolithic Revolution** commenced in the Middle East. The term is used as a shorthand to describe the period when humans first began to domesticate animals and to plant, cultivate and improve edible grasses and roots. In other words, humans started to modify the environment rather than merely adapting to it, so that the possibility of establishing permanent settlements came about (Carol, 1964). A seemingly universal attribute of Neolithic cultures was the cultivation of wheat and barley, whilst the notable development in the realm of material culture was the manufacture of pots. These and a whole series of subsequent inventions and discoveries, such as the plough, the wheeled cart, the sailing boat and the chemical processes of smelting, equipped humans for urban life (Childe, 1951).

The emphasis must, however, be placed on the word 'equipped'. Naturally the transition from food gathering to food producing and the attendant increase in food production, both per head of the population and per unit of land, must have increased greatly the carrying capacity of the land. But these developments alone cannot explain the emergence of the first urban settlements. Indeed, with respect to Egypt, Mesopotamia and the Indus Valley, there is much evidence that initially these changes did not lead to growth in the size of individual settlements, but rather to an increasing number of small self-sufficient villages. Put simply, the Neolithic Revolution could have led to more settlements and a higher density of population without instigating urbanisation itself.

It was Childe (1950) who argued that the direct link between Neolithic villages and later city life was the production of a **food surplus**. Hence, it was stated that the first cities were the outcome of favourable hydrological and environmental circumstances, such as the existence of periodically flooded river valleys in the 'fertile crescent' and elsewhere. However, a food surplus can only be seen as a *necessary but not sufficient precondition* to urbanisation. We are forced to ask why agriculturalists would be prepared to give up some of their produce in order to support directly a non-producing ruling elite. Clearly, for this first occurrence of differential power and occupational specialisation, further substantial catalysts were required.

The move away from an entirely ecological–hydrological explanation of the first cities owes much to Adams (1966). Against all conventional wisdoms, Adams argued that **social factors** were paramount in the creation of the first cities. More specifically, Adams maintained that it was social change that precipitated agricultural and technical change and not the other way around. This argument has since gained general currency and has been developed by several scholars (Johnston, 1977).

Thus, both the military and the religious theories of urban genesis argue that elite groups emerged and extracted the food surplus as a tribute, in return for either spiritual and/or physical protection and security. Such ideas suggest possible reasons why the economically and socially exploited group might have been prepared to give up part of its surplus product. Further, it is salient that the settlements in all seven areas of primary urban generation were indeed associated with religio-military temples and forts. The role of religion in urban generation has been most strongly argued by Wheatley (1971). Warfare and military explanations of urban origins seem credible, although they are probably best viewed as factors that intensified urbanism once it had actually been established. We reach an important conclusion for the consideration of urbanisation at any time or place – namely that urban living has always been intimately associated with the disposition of power and social differentiation within society.

These socially oriented theories do provide suggestions as to how egalitarian societies became both occupationally specialised and socially differentiated, so that the transition from the reciprocal stage to that of rank redistribution could occur. Other commentators have advanced the economic thesis that trade and marketing functions were the prime progenitors of the first cities, Jane Jacobs (1969) being notable among them. But as Carter (1977) concludes, as with military explanations, although trade was clearly an important element in the intensification of early cities, it is probably best regarded as a by-product of urbanism, not as its creator. This view is also exemplified by Johnston (1980) when he argues that part of the economic benefits of trade were used as further rewards for the exploited segment if, and when, scepticism grew about the benefits of the spiritual or temporal security afforded by urban living.

Although it can be asserted that the prime factors in urban genesis were social, it seems sensible to conclude that *no single* causal factors could have been operative. This view is well demonstrated by Sjoberg (1960, 1965), who argues that the course of pre-industrial urban development was intimately related to three pre-conditions: a relatively favourable ecological base, allied to a relatively advanced technology, in turn linked with a new type of social organisation, in the form of the emergence of political and economic power. Such impetuses would inevitably become cumulative and mutually reinforcing and would subsequently be strengthened by military and trade functions. Similarly, it can be ventured that the self-sustaining nature of the urban process would have been bolstered by increasing social heterogeneity. Once the advantages of wealth, power and status had become vested with particular groups, it is likely that they would seek not only to preserve, but to strengthen, the existing system, an argument that has direct relevance to contemporary urbanisation in the developing world, and which is developed in Chapter 4. Presumably, the continued development of urban living was smoothest where the standard of living of the exploited group increased in absolute terms, despite the mounting relative disparity, a further argument that holds salience for present-day urban conditions. Thus, the relationship between the economically powerful and the subject class under urbanisation was simultaneously both exploitative and protective.

The intention of the account so far has been to stress the processes involved in urbanisation, for questions concerning the precise descriptive details of urban origins abound. For example, the issue of the size of such centres is frequently raised and it is asked whether these settlements were truly urban in character. In this respect, estimates vary. With regard to Mesopotamian cities, Sjoberg talks of populations from 5000 to 10,000, and Childe of 7000–20,000. However, Childe (1950) surveyed the features of all known cities of antiquity and concluded that they were all truly urban. His judgement was not based on their size alone, but was related to the density, occupational specialisation and class structure of these cities, along with their role in the collection, storage and distribution of the surplus, as well as their monumental structure and artistic achievements.

Finally, there is as yet no universal acceptance of the designation of the seven areas mentioned as being the only possible sites of primary urban generation. In particular, excavations on the Anatolian plateau in Turkey at Çatal Höyük show evidence of ceremonial shrines and graveyards dating back to 6500 BC. Thus, it has been suggested that prior to 6000 BC, some three or four millennia before the famous cities of the Middle East, Çatal Höyük was a fully fledged and thriving urban centre (Mellaart, 1964, 1967). These facts add evidence to the process explanations offered in the present account, for this upland area did not enjoy any of the advantages of the riverine valleys, indicating that urban living was not triggered by favourable ecological circumstances and a food surplus.

Opening up the debate: cities and social surplus product

The important lesson to be drawn from the review of urban origins is that the character and form of urbanisation cannot be divorced from the nature of development itself. It follows therefore, that changes in socio-political organisation are required to effect urban change. Further, it is manifestly the case that cities have always served, and been associated with, elite groups, whether religious and military as in Mesopotamia, an expatriate elite as in the mercantile – colonial city, or the corporate interests of present-day multinational companies.

Associated with this is the central fact that cities have always been intimately associated with the generation of a **surplus**. Indeed, it can be argued that without a **social surplus product**, cities cannot exist (Plate 2.1). It is

Plate 2.1   Cities and surpluses: informal sector transport in Delhi as part of surplus redistribution (photo: Rob Potter).

axiomatic that views of the processes involved will be coloured by political predispositions. The definition of social surplus product is an excess of production over biological need, so that some members of the society are released from the need to produce.

The issue has been treated from a Marxist perspective by Harvey (1973), who follows the view that 'cities are formed through the geographic concentration of a social surplus product' (Harvey, 1973: 216). It may be argued that a surplus can take two quite distinct forms. In the first, it is an amount of material product that is set aside to provide for improvements in human welfare. This may be regarded as the *communist form of a surplus*. As Harvey and others have noted, a social surplus is required to advance a socialist society, but it is argued that this does not have to be, and indeed should not be, class-based in character, but rather should be used for socially oriented or communally defined purposes. Thus, Harvey (1973) notes that whilst a surplus is required in a socialist society, there is no *a priori* reason why it should be spatially concentrated. However, Harvey goes on to argue that as investment may well be more efficiently deployed in concentrated form due to the operation of economies of scale and agglomeration, some type of urbanisation may well be acceptable. But in so far as the surplus is distributed for the use of the population in general, too great a degree of geographic concentration should be avoided. As an example, Harvey cites the case of Cuba, where following the revolution of 1959, a conscious attempt was made to disperse medical facilities away from Havana, an example which is reviewed in Chapter 4 (Plate 2.2).

Plate 2.2 Havana, Cuba: concentrated social surplus characterises urbanity under socialism as well as capitalism (photo: Rob Potter).

The second kind of surplus is described as an estranged or alienated version of the first. Simply stated, it is a quantity of material resources and product that is appropriated for one segment of society at the expense of others. Harvey argues that it is this form of surplus that has characterised Neolithic urbanism, rural–urban flows and contemporary urbanism under capitalism. This gives rise to the argument that profits are the vital signal to the economy and that inequalities are a vital pre-requisite for economic growth and efficiency. This is, of course, the starting point for the traditional–classical view of economic development.

Under a capitalist formulation, it is posited that surplus value in the form of profits is invested in realising further profits. Hence, Harvey (1973) points to the monumental architecture, conspicuous consumption and need creation associated with the contemporary city. Some argue that under capitalism, urbanisation is legitimised by its contribution to gross national product. Broadly, it is asserted that investment should be concentrated in the most profitable areas in order to maximise growth. Thus, advocates of the free-market system argue that surpluses in the form of profits signal the demands of consumers to producers. It is a simple step to state that the city is a mechanism of economic efficiency and growth, and thus eventually, increasing prosperity for all. This thesis relates directly to the traditional idea that cities are intimately associated with economic development (see, for example, Gottmann, 1983). Certainly, the city has often been regarded as a centre of mixing, discovery and innovation, benefiting from scale and agglomeration economies. This is epitomised by Lampard's (1955: 92) statement that the 'modern city is a mode of social organisation which furthers efficiency in economic activity'. This, of course, is the springboard for traditional **neo-classical–capitalist** models of development as well as recent **neo-liberal reformulations.**

These two contrasting political views on the role of the city are in agreement in so far as they both recognise that cities are associated with the generation of surpluses and wealth. Where they differ is in their assessment of the acceptability of the final outcome. This really boils down to an evaluation of the social, economic and ethical–moral costs and benefits and the net outcome of urban agglomeration. The issue was raised early on by Hoselitz (1955), who suggested in somewhat simplistic terms that cities can be classified as either 'generative' or 'parasitic', according to whether or not they stimulate economic growth in their wider regions. Hoselitz conceded that most early colonial settlements were parasitic, that is they extracted surplus value from surrounding regions. But he maintained that they later generally became generative by virtue of widening out economic development over a more extensive area and an increasing proportion of the population. Whether, in fact, such a process occurs, and if so, over what time scale, is a critical issue which lies at the heart of the development studies literature. A major point which needs to be addressed, however, is that as noted in Chapter 1, cities in the contemporary developing world are more an expression of a lack of economic development, rather than a sign of it. Similarly, in Western societies, the demise of the inner city and industrial restructuring mean that the future of the city as we know it looks increasingly uncertain.

Whether cities are an evil, a necessary evil or the greatest of human achieve-ments is a central, although contentious, question. However, what is beyond dispute is that cities have always been closely associated with the generation of increasing inequalities. Facets of both global and national inequalities underlie patterns of present-day Third World urban development. Thus, Gilbert and Gugler (1992: 11) point out that in 'an unequal world, therefore, it is not surprising that cities should also be unequal'. The theme of spatial polarisation and inequality is specifically addressed in this and the next chapter. The range of theories and conceptual models which have been developed in order to examine patterns and processes of growth will be reviewed in the second half of the present chapter. In effect, each of these models may be regarded as dealing with the nature of surplus redistribution. In Chapter 3, the focus is on the structure of urban systems at the international, national and regional levels.

Cities and surpluses in developing countries: theoretical and conceptual approaches

Broadly speaking, it is posited here that four major approaches to the examina-tion of the relationships between cities and the generation of social surplus product can be recognised: the classical–traditional approach, the historical approach, the radical–political economy–dependency approach, and finally what may be referred to as the bottom-up and postmodern approach. Each of these approaches may be regarded as expressing a particular ideological stand-point, and can also be identified by virtue of having occupied the centre stage of the development debate at a particular point in time. However, each ap-proach still has currency. Hence, in development theory, left-of-centre socialist views may be more popular than neo-classical formulations, but in practice the 1980s and 1990s have seen the implementation of neo-liberal interpretation of classical approaches, stressing public sector cutbacks as a part of structural adjustment programmes. Notwithstanding this caveat, the account which fol-lows uses these four divisions to overview the leading theories which have been produced to explain the role of cities in the development process.

Classical–traditional approaches

The classical approach to the study of development derives from neo-classical economics and totally dominated thinking about development for a period close on 40 years. Summarised in simple terms – and some commentators would argue that such approaches rarely attempted anything more sophistic-ated – these approaches basically argue that developing countries are charac-terised by a **dualistic structure**. The dualism exists between a traditional, indigenous, underdeveloped sector on the one hand, and a modern, developed and westernised one on the other. Seminal works include those of Meier and Baldwin (1957), Prebish (1950), Perloff and Wingo (1961), Schultz (1953), Perroux (1950, 1955), Myrdal (1957) and Hirschman (1958).

The general economic development model of the American economist A.O. Hirschman forms a convenient springboard for discussion of the approach. Hirschman (1958), in his volume *The Strategy of Economic Development*, advanced a notably optimistic view in presenting the neo-classical position (see Hansen, 1981). Specifically, Hirschman argued that polarisation must be regarded as an inevitable characteristic of the early stages of economic development. This represents the direct advocacy of a basically unbalanced economic growth strategy, whereby investment is concentrated in a few key sectors of the economy. It is envisaged that the growth of these sectors will create demand for the other sectors of the economy, so that a 'chain of disequilibria will lead to growth'. The corollary of sectorally unbalanced growth is geographically uneven development, and Hirschman specifically cited Perroux's (1955) concept of the natural growth pole.

The forces of concentration were collectively referred to by Hirschman as '**polarisation**'. The crucial argument, however, was that eventually development in the core will lead to the '**trickling down**' of growth to backward regions. These trickle-down effects were seen by Hirschman as an inevitable and spontaneous process. Thus, the clear policy implication of Hirschman's thesis is that governments should not intervene to reduce inequalities, for at some juncture in the future, the search for profits will promote the spontaneous spin-off of growth, inducing industries to backward regions. The process whereby spatial polarisation trends give way to spatial dispersion out from the core to the backward regions has subsequently come to be known as the point of '**polarisation reversal**' (Richardson, 1977, 1980).

The true significance of these ideas concerning polarised development extends beyond their use as a basis for understanding the historical processes of urban–industrial change, for in the 1950s and 1960s they came to represent an explicit framework for regional development policy (see Friedmann and Weaver, 1979). Thus the doctrine of unequal growth gained both positive and normative currency in the first post-war decade and the path to growth was actively pursued via urban-based industrial growth. The policies of non-intervention, enhancing natural growth centres, and creating new induced sub-cores became the order of the day. As Friedmann and Weaver (1979: 93) observe, the 'argument boiled down to this: inequality was efficient for growth, equality was inefficient', so that, 'given these assumptions about economic growth, the expansion of manufacturing was regarded as the major propulsive force'.

Hirschman's ideas can be seen as part of wider **modernisation theory**. The paradigm was grounded in the view that the gaps in development which exist between the developed and developing countries can gradually be overcome on an imitative basis, and thereby developing countries would inexorably come to resemble developed countries (Hettne, 1994). The modernisation thesis was largely developed in the field of political science, but was picked up by a group of geographers in the late 1960s (see, for instance, Soja, 1968, 1974; Gould, 1970; Riddell, 1970). In such work, sets of indices which were held to reflect modernisation were mapped and/or subjected to multivariate statistical analysis to reveal what was regarded as the '**modernisation surface**'. For example, using such an approach, Gould (1970) examined the so-called modernisation surface of Tanzania.

A classic paper written in the mould was by Leinbach (1972), who investigated the modernisation surface in Malaya between 1895 and 1969, using indicators such as the number of hospitals and schools per head of the population, together with the incidence of postal and telegraph facilities and road and rail densities. The modernisation approach did serve to emphasise that core urban areas and the transport corridors running between them are the focus of dynamic change (Leinbach, 1972), although from a critical perspective, Friedmann and Weaver (1979: 120) observed that the approach only succeeded in 'mapping the penetration of neo-colonial capitalism'.

The hallmark of such work was that it posited that modernisation is basically a temporal–spatial process. In such a formulation, underdevelopment is seen as something which can be overcome, principally by the spatial diffusion of **modernity**. A number of studies argued that growth occurs within the settlement system from the largest urban places to the smallest in a basically hierarchical sequence. Foremost among the proponents of such a view was Hudson (1969) who applied the classic ideas of Hagerstrand (1953) to the settlement or central place system. Hudson argued that innovations can travel through the settlement system by a process of contagious spread, where there is a neighbourhood or regional effect of clustered growth. This was close to Schumpeter's (1911) general economic theory, in which he argued that the essence of development is a volume of innovations. Opportunities tend to occur in waves which surge after an initial innovation. Thus, Schumpeter argued that development tends to be 'jerky' and to occur in 'swarms'.

Alternatively, Hudson noted that diffusion can also occur downwards through the settlement system in a progressive manner, the point of introduction being the largest city. Pedersen (1970) argued the case for a strictly hierarchical process of innovation diffusion of this kind, an assertion which seemed to be substantiated by some empirical–historical studies carried out in advanced capitalist societies such as the United States (Borchert, 1967) and England and Wales (Robson, 1973). Pedersen drew a very important distinction between domestic and entrepreneurial innovations, the latter being the instrument of urban growth, rather than the former. In another frequently cited paper of the time, Berry (1972) also argued strongly in favour of a **hierarchical diffusion process** of growth-inducing innovations, this basically being seen as the result of the sequential market searching procedures of firms, along with imitation effects. But notably, Berry's analysis was based on the diffusion of domestic as opposed to entrepreneurial innovations, namely of television stations and receivers.

All of these approaches, involving unequal and uneven growth, modernisation, the diffusion of innovations and hierarchic patterns of change, may be grouped together and regarded as constituting the **top-down paradigm of development** (Stöhr and Taylor, 1981), which advocated the establishment of strong urban–industrial nodes as the basis of self-sustained growth. Such an approach is premised on the occurrence of strong trickle-down effects, by means of which it is believed that modernisation will inexorably spread from urban to rural areas. Thereby, such models, including Rostow's (1960) classic *The Stages of Economic Growth*, see cities as engines of growth and development. Rostow envisaged that there were five stages through which all developed countries have to pass: the traditional society, the pre-conditions to take-off, take-off, the road to maturity, and the age of mass consumption. His stage model encapsulates faith in the capitalist system, as expressed by the subtitle of Rostow's work: 'a Non-Communist Manifesto'. Such formulations place absolute faith in the existence of a linear and rationalistic path to development, based on western positivism and science, and the possibility that all nations can follow this in an unconstrained manner. Modernism was very much an urban phenomenon from 1850 onwards (Harvey, 1989). Universal or high modernism became hegemonic after 1945. Thus, the top-down approach was strongly associated with the 1950s, through to the early 1970s.

Historical approaches: realism dawns?

In contrast to Hirschman, the Swedish economist Gunnar Myrdal (1957), although writing at much the same time, took a noticeably more pessimistic view, maintaining that capitalist development is inevitably marked by deepening regional and personal income and welfare inequalities. Myrdal followed the arguments of the vicious circle of poverty in presenting his theory of 'cumulative causation'. Thereby, it was argued that once differential growth occurs, thereafter internal and external economies of scale will perpetuate the pattern. This is the outcome of the '**backwash**' effect whereby population migrations, trade and capital movements all come to focus on the key growth points of the economy. Increasing demand, associated with multiplier effects and the existence of social facilities, also serve to enhance the core region. Whilst '**spread**' effects will undoubtedly occur, principally via the increased market for the agricultural products and raw materials of the periphery, Myrdal basically concluded that given unrestrained free-market forces, these would in no way match the backwash effects. Myrdal's thesis leads to the advocacy of strong state policy in order to counteract what is seen as the normal tendency of the capitalist system to foster increasing regional inequalities (see Hollier, 1988).

The view that, without intervention, social surplus product and development are both likely to become increasingly polarised in transitional societies was taken up and developed by a number of scholars towards the end of the 1960s and the beginning of the 1970s. In so doing they ran counter to the conventional wisdom of the time. Such works were based mainly on studies which encompassed an historical dimension. Perhaps the best-known example is provided by the American planner John Friedmann's (1966) **core–periphery model**. From a purely theoretical perspective, Friedmann's central contention was that 'where economic growth is sustained over long time periods, its incidence works toward a progressive integration of the space economy' (Friedmann, 1966: 35). This process is made clear in the much-reproduced four-stage ideal–typical sequence of development, which is shown in Figure 2.1.

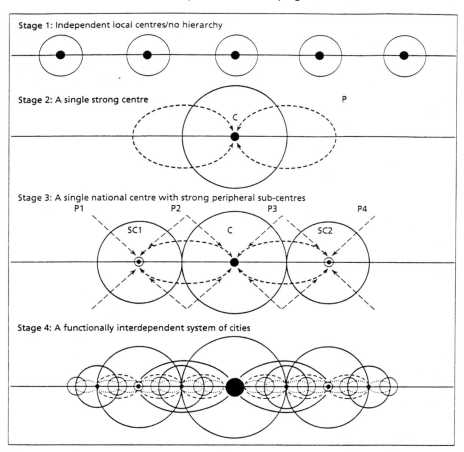

Stage 1: Independent local centres/no hierarchy

Stage 2: A single strong centre

Stage 3: A single national centre with strong peripheral sub-centres

Stage 4: A functionally interdependent system of cities

Figure 2.1   A summary of Friedmann's core–periphery model (adapted from Friedmann, 1966).

The first stage, that of *independent local centres with no hierarchy*, represents the pre-industrial stage and is associated with a series of isolated self-sufficient local economies. There is no social surplus product to be concentrated in space and an even and essentially stable pattern is the result (Figure 2.1).

In stage 2, *a single strong centre*, it is posited that as the result of some form of 'external disruption' – a euphemism for colonialism – the former stability is replaced by dynamic change. Growth is envisaged to occur rapidly in one main region and urban primacy is the spatial outcome. Social surplus product is strongly concentrated. The centre (C) feeds on the rest of the nation, so that the extensive periphery (P) is drained. Advantage tends to accrue to a small elite of urban consumers, who are located at the centre. However, Friedmann regarded this stage as inherently unstable (Figure 2.1).

The outcome of this instability is the development of a *single national centre with strong peripheral sub-centres* (Figure 2.1). Over time, the simple centre–periphery pattern is progressively transformed to a multinuclear one. Sub-cores develop (SC1, SC2) leaving a series of inter-metropolitan peripheries (P1 to P4). This is the theoretical representation of the point of polarisation reversal when social surplus product starts to be concentrated in parts of the former periphery, albeit on a highly concentrated basis.

The fourth and final stage, which sees the development of a *functionally interdependent system of cities*, was described by Friedmann as 'organised complexity' and is one where progressive national integration continues, eventually witnessing the total absorption of the inter-metropolitan peripheries. A smooth progression of cities by size is envisaged as the outcome (Figure 2.1).

The first two stages of the core–periphery model describe directly the history of the majority of most developing countries. Indeed, it often appears not to be appreciated that the line along which the small independent communities are drawn in stage 1 represents the coastline. The occurrence of uneven growth and urban concentration in the early stages of growth is seen as being the direct outcome of exogenic forces. Thus, Friedmann commented that the core–periphery relationship is essentially a colonial one, his work having been based on the history of regional development in Venezuela.

The principal idea behind the centre–periphery framework is that early on, factors of production will be displaced from the periphery to the centre, where marginal productivities are higher. Thus, at an early stage of development, nothing succeeds like success. However, the crucial change, of course, is the transition between the second and third stages, that is where the system tends towards equilibrium and equalisation. Friedmann's model is one which suggests that in theory, economic development will ultimately lead to the convergence of regional incomes and welfare differentials.

But at the very same time as he was presenting the simplified model as a template, Friedmann observed that in reality there was evidence of persistent disequilibrium. Thus, in a statement which appeared alongside the model, Friedmann (1966: 14) observed that there was 'a major difficulty with the equilibrium model: historical evidence does not support it'. Despite this damning caveat many authors have represented the model as a statement of invariant truth, ignoring Friedmann's warning that 'disequilibrium is built into transitional societies from the start' (Friedmann, 1966: 14). Effectively, Friedmann was maintaining that without state intervention, the transition from stage 2 to stage 3 of the model will not occur in developing societies, and in this respect, he was in agreement with Myrdal's prescriptions that social surplus product will become ever more concentrated in space.

Writing just a few years after the appearance of Friedmann's much-cited model, an American geographer, Jay E. Vance (1970), noted that it was with the development of mercantile societies from the fifteenth century onward that settlement systems started to evolve along more complex lines. The main development, as noted in Chapter 1, came with colonialism, for continued economic growth required greater land resources. Frequently, that was initially met by local colonial expansion via trading expeditions. By the seventeenth and eighteenth centuries, however, this need was increasingly fulfilled by distant colonialism, that is the trans-oceanic version of local colonialism. The implications of these historical developments have been well summarised by Vance (1970: 148):

The vigorous mercantile entrepreneur of the seventeenth and eighteenth century had to turn outward from Europe because the long history of parochial trade and the confining honeycomb of Christaller cells that had grown up with feudalism left little scope there for his activity. With overseas development, for the first time the merchant faced an unorganised land wherein the designs he established furnished the geography of wholesale-trade location. By contrast, in a central-place situation (such as that affecting much of Europe and the Orient), to introduce wholesale trade meant to conform to a settlement pattern that was premercantile.

The development of mercantilism has already been summarised in Chapter 1. During this period, ports came to dominate the evolving urban systems of both the colony and the colonial power. In the colony, once established, ports acted as **gateways** to the interior lands. Subsequently, evolutionary changes occurred that first saw increasing spatial concentration at certain nodes, and later, lateral interconnection of the coastal gateways and the establishment of new inland regions for expansion. The settlement pattern of the homeland also underwent considerable change, for social surplus product flowed into the capital city and the principal ports, thus serving to strengthen considerably their position in the urban system.

These historical facets of trade articulation led Vance (1970) to suggest what amounted to an entirely new model of urban settlement evolution, one that was firmly based on history. This is referred to as the **mercantile model**, and its main features are summarised in Figure 2.2 (overleaf). The model is summarised in five stages. In each of these the colony is shown on the left of the figure, and the colonial power on the right.

The first stage represents the *initial search phase of mercantilism*, involving the search for economic information on the part of the prospective colonising power. The second stage sees the *testing of productivity and the harvest of natural storage*, with the periodic harvesting of staples such as fish, furs and timber. However, no permanent settlement is established in the colony. The *planting of settlers who produce staples and consume the manufactures of the home country* represents the third stage. The settlement system of the colony is established via a point of attachment. The developing symbiotic relationship between the colony and the colonial power is witnessed by a sharp reduction in the effective distance separating them. The major port in the homeland becomes pre-eminent. The fourth stage is characterised by the *introduction of internal trade and manufacture in the colony*. At this juncture, penetration occurs inland from the major gateways in the colony, based on staple production. There is rapid growth of manufacturing in the homeland to supply both the overseas and home markets. Ports continue to increase in significance. The fifth and final stage sees the *establishment of a mercantile settlement pattern* with central place infilling occurring in the colony; and the emergence of a central place-type settlement system with a mercantile overlay in the homeland.

The mercantile model stresses the historical–evolutionary viewpoint in examining the development of national settlement systems. The framework offers what Vance sees as an alternative and more realistic picture of settlement structure, based on the fact that in the seventeenth and eighteenth centuries,

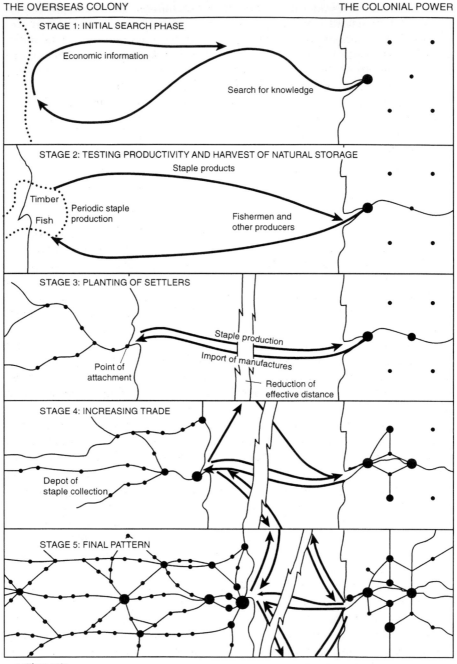

THE OVERSEAS COLONY   THE COLONIAL POWER

STAGE 1: INITIAL SEARCH PHASE

Economic information

Search for knowledge

STAGE 2: TESTING PRODUCTIVITY AND HARVEST OF NATURAL STORAGE

Staple products

Timber

Periodic staple
production

Fish

Fishermen and
other producers

STAGE 3: PLANTING OF SETTLERS

Staple production

Import of manufactures

Point of
attachment

Reduction of
effective distance

STAGE 4: INCREASING TRADE

Depot of
staple collection

STAGE 5: FINAL PATTERN

• settlements

Figure 2.2   A simplified representation of Vance's mercantile model (source: Potter, 1992b).

mercantile entrepreneurs turned outward from Europe. Hence, the source of change is external to developing countries, a theme which is continued in Chapter 3 in a contemporary context. In contrast, the development of settlement patterns and systems of central places in the developed world was based on endogenic principles of local demand, thereby rendering what is essentially a closed settlement system (Christaller, 1933; Lösch, 1940).

The hallmark of the mercantile model is the remarkable linearity of settlement patterns, first along coasts (especially in colonies), and secondly along the routes which developed between the coastal points of attachment and the staple-producing interiors. These two alignments are also given direct expression in Taaffe, Morrill and Gould's (1963) model of transport expansion in less developed countries, based on the transport histories of Brazil, Malaya, East Africa, Nigeria and Ghana.

In plantation economies such as those of the Caribbean, a local historical variant of the mercantile settlement system is provided by the **plantopolis model**. A simplified representation of this is shown in Figure 2.3 (overleaf). The first two stages are based on Rojas (1989), although the graphical depiction of the sequence and its extension to the modern era have been effected by Potter (1995). In stage 1, *plantopolis* (1750–), the plantations formed self-contained bases of the settlement pattern, such that only one main town performing trade, service and political control functions was required. Following *emancipation* (1833–), small, marginal farming communities, which were clustered around the plantations practising subsistence agriculture and supplying labour to the plantations, added a third layer to the settlement system (stage 2). The distribution of these, of course, would vary according to physical and agricultural conditions.

The figure suggests that, in the Caribbean, the *modern era* (1950–) has witnessed the extension of this highly polarised pattern of development, as shown in stage 3 of Figure 2.3. The emphasis should be placed on extension, for this may not in all cases amount to intensification *per se*. This has come about largely as the result of industrialisation and tourism being taken as the twin paths to development. In 1976, Augelli and West (1976: 120) commented on what they regarded as the disproportionate concentration of wealth, power and social status in the chief urban centres of the West Indies. As shown in Figure 2.3, such spatial inequality is sustained by strong symbiotic flows between town and country.

The virtues of the mercantile and plantopolis models are many. Principally, they serve to stress that the development of settlement systems in most developing countries amounts to a form of **dependent urbanisation**. Certainly we are reminded that the high degree of urban primacy and the littoral orientation of settlement fabrics in Africa, Asia, South America and the Caribbean are all the direct product of colonialism, not accidental happenings or aberrant cases, hence the comment that modernisation surfaces merely chart neo-colonial penetration. According to this framework, ports and other urban settlements became the focus of economic activity and of the social surplus which accrued. A similar but somewhat less overriding spatial concentration also applies to the

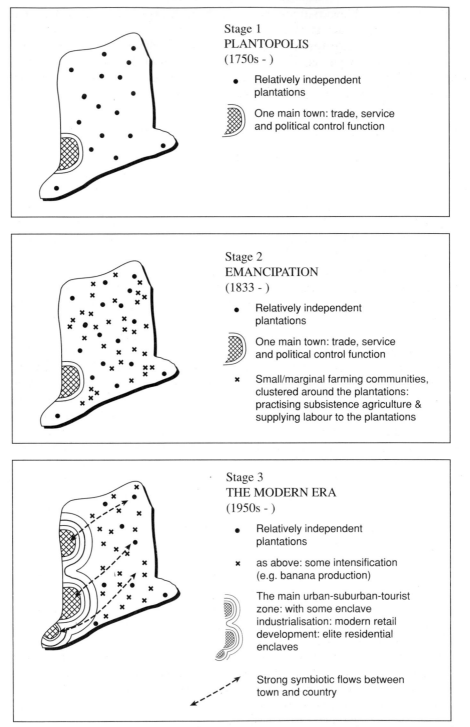

Figure 2.3   An extension of the plantopolis model into the 'modern era' (source: Potter, 1995).

colonial power. Hence, a pattern of spatially unequal or polarised growth emerged strongly several hundred years ago with the strengthening of this symbiotic relationship between colony and colonial power. The overall suggestion is that due to the requirements of the international economy, far greater levels of inequality and spatial concentration of social surplus product are produced than may be socially and morally desirable.

Radical political economy–dependency approaches

The mercantile and plantopolis models can be seen as graphical depictions of the outcome of the interdependent development of the globe since the 1400s. The '**dependencia**' or **dependency school** specifically took this theme up as a rebuttal of the modernisation paradigm. Dependency theory also developed as a voice from the Third World, rather than as a reflection of Eurocentric development thinking (Hettne, 1994), reflecting the writings of Latin American and Caribbean structuralists such as Beckford, Prebish and Cardoso.

The dependency approach is particularly associated with the work of Andre Gunder Frank, a Chicago-trained economist, who underwent a rapid and thorough radical conversion. In *Capitalism and Underdevelopment in Latin America*, Frank (1967) maintained that development and underdevelopment are opposite sides of the same coin, and that both are the necessary outcome and manifestation of the contradictions of the capitalist system of development. The thesis presented by Frank was devastatingly simple. He argued that the condition of developing countries is not the outcome of inertia, misfortune, chance, climatic conditions or whatever, but a reflection of the manner of their incorporation into the world capitalist system. Viewed in this manner, so-called 'underdevelopment', and associated dualism, were not a negative, void or failure, but the direct outcome and reciprocal of development elsewhere.

The expression 'the development of underdevelopment' (Frank, 1966) has come to be employed as a shorthand description of the approach, which has a strong graphical tie-in with the mercantile and plantopolis models. Quite simply, if the development of large tracts of the earth's surface has been dependent upon metropolitan cores, referred to as **metropoles**, then the development of cities has also been dependent, principally upon the articulation of capital and the accumulation of surplus value (see Figure 2.4, overleaf). The process has operated both internationally and internally within countries. Viewed in this light, so-called 'backwardness' results from integration at the bottom of the hierarchy of dependence. Frank argued that the more such '**satellites**' are associated with the metropoles, the more they are held back, and not the other way around. In this connection Frank specifically cited the instances of northeast Brazil and the West Indies as regions where processes of internal transformation had been rendered impossible due to such close contact. There was, in fact, a strong Caribbean dependency school writing in the early 1970s, with George Beckford (1972) in an important book, *Persistent Poverty*, examining the 'plantation economy' as a local variant of dependency (see also Girvan, 1973; Cumper, 1974).

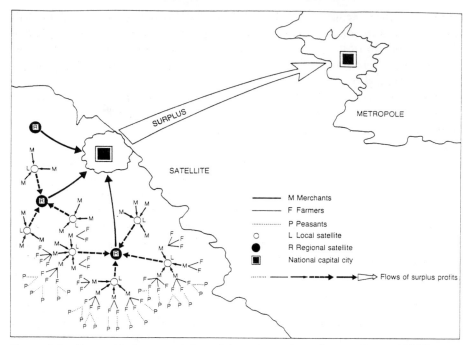

Figure 2.4 Geographical representation of dependency theory (source: Potter, 1992b).

Dependency theory represents an holistic view because it describes a chain of dependent relations which has grown since the establishment of capitalism as the dominant world system, so that its expansion is regarded as co-terminous with colonialism and underdevelopment. The chain of exploitative relations witnesses the extraction and transmission of surplus value via a process of **unequal exchange**, extending from the peasant, through the market town, regional centre, national capital, to the international metropole, as shown in Figure 2.4. The **terms of trade** have always worked in favour of the next higher level in the chain, so that social surplus value is progressively concentrated (see also Harvey, 1973; Castells, 1977). By such means, dependency theorists argue, the dominant capitalist powers, such as the United Kingdom and then the United States, encouraged the transformation of political and economic structures, in order to serve their interests. According to this view, colonial territories were organised to produce primary products at minimal cost, whilst at the same time becoming an increasing market for industrial products. Inexorably, social surplus value was siphoned off from poor to rich regions, and from the developing world to the developed. Such a process is an international extension of that which guided the development of the very first cities.

The chief criticism of dependency theory is that it is economistic, seeing all as the outcome of a form of economic determinism, conforming with what Armstrong and McGee (1985: 38–39) have described as the 'impersonal, even mechanical analysis of structuralism'. Further, the theory only appears to deal

with class structure and other factors internal to a given nation, as the outcome of the economic processes described. Another point of contention is that
dependency theory suggests that socialist countries can only advance their lots
by de-linking from the global economy, whilst the capitalist world system is
busily becoming more global.

Wallerstein (1974, 1980) attempted to get around some of these issues,
including the internal–external agency debate, by stressing the existence of a
world system (see also Taylor, 1986). The essential point is that Wallerstein
distinguishes not only between the **core nations** which became the leading
industrial producers, and the **peripheral states** which were maintained as agricultural providers, but also identifies the **semi-peripheries.** These play a key
role, for these intermediate states are strongly ambitious in competing for core
status by means of increasing their importance as industrial producers, relative
to their standing as agricultural suppliers. Within the world system since the
sixteenth century, there have been cyclical periods of expansion, contraction,
crisis and change. Hence, it is envisaged that the fate of a particular nation is
not entirely externally driven, but depends on the internal manner in which
external forces have been accommodated.

Frank's ideas are seen by many as being near to the orthodox Marxist view
that the advanced capitalist world at once both exploited and kept the Third
World underdeveloped. Whilst many would undoubtedly refute this view,
regarding it as being extreme, if one can clear away the moral outrage, elements of the analysis, even if in world systems form, are likely to provide food
for thought for those interpreting urban patterns and processes in the contemporary developing world. As already observed, certainly the graphical representation of pure dependency theory exhibits many parallels with the spatial
outcomes of the core–periphery, mercantile and plantopolis models of settlement development and structure.

The bottom–up paradigm and postmodern approaches to development and change

From the mid-1970s, a growing critique of top-down policies, and especially
growth pole policy, argued that they merely replaced concentration at one
point with concentrated deconcentration at a limited number of new localities.
The suggestion that there is only one, linear path to development, and that
development is the same thing as economic growth, were accepted wisdoms
which came at long last to be challenged. Liberal and radical commentators
averred that top-down approaches to development were acting as the hand-
maidens of transnational capital.

The latter point was emphasised in an important book which sought to
consider the evolution of regional planning and practice, under the title *Territory and Function.* This volume was written by John Friedmann and Clyde
Weaver (1979). In it, the authors argued that development theory and practice
up to that time had been dominated by purely functional concerns relating to

efficiency and modernity, with all too little consideration being accorded to the needs of particular territories, and to the territorial bases of development and change. Since 1975, a major new paradigm has come to the fore, which involves stronger emphasis being placed on rural-based strategies of development. As a whole, this approach is described as **development from below**. Other terms used to describe the paradigm include agropolitan development, grassroots development and urban-based rural development.

The principal idea is that basic needs must first be met within particular territories. In the purest form, it is argued that this can only be achieved by nations becoming more reliant on local resources, the communalisation of productive wealth, and closing up to outside forces of change. This is known as **selective regional/territorial closure.** In simpler words, it is argued that Third World countries should try to reduce their involvement in processes of unequal exchange. The only way round the problem is to increase **self-sufficiency** and **reliance**. It is envisaged, however, that later, the economy can be diversified and non-agricultural activities introduced. But it is argued that in these circumstances, urban locations are no longer mandatory, and cities can in this sense be based on agriculture. Thus, Friedmann and Weaver (1979: 200) comment that 'large cities will lose their present overwhelming advantage'. Clearly, the approach is inspired by, if not entirely based on, socialist principles. Classic examples of the enactment of bottom-up paths to development have been China, Cuba, Grenada and Tanzania, and the experiences of these and other countries will be evaluated in Chapter 4.

Walter Stöhr (1981) provides an informative overview of the development from below strategy. In particular, his account stresses that there is no single recipe for such strategies, as there is for those from above. Development from below needs to be closely related to specific socio-cultural, historical and institutional conditions. Simply stated, development should be based on territorial units and should endeavour to mobilise their indigenous natural and human resources.

Bottom-up strategies are varied, with alternative paths to development being stressed. They share the characteristic of arguing that social surplus product should not be concentrated at each higher level of the settlement and social systems, but should focus on the needs of the lower echelons of these orders. It is this characteristic which gives rise to the term 'bottom-up', for such strategies are, in fact, often enacted by strong state control and direction from the political 'centre'.

Such trends in development can be linked with the idea that we are entering the **postmodern age**. At a straightforward level, postmodernity involves moving away from an era dominated by notions of modernisation. It is, therefore, intimately associated with development theory and practice. It involves the rejection of modernism and a return to premodern and vernacular forms, as well as the creation of distinctly new postmodern forms (see Urry, 1990b; Harvey, 1989). There is much in the argument that the whole ethos of the modern period privileged the metropolitan over the provinces, the developed over the developing worlds, North America over the Pacific Rim, the professional over the general populace, and men over women. In contrast, the postmodern world potentially involves a diversity of approaches, which may serve to empower 'other' alternative voices and cultures. A strong emphasis on bottom-up, non-hierarchical growth strategies, which endeavour to get away from international sameness, can be seen as part and parcel of the approach. The accent can potentially be placed on growth in smaller places rather than bigger, in the periphery and not the core.

However, although postmodernism may in certain respects be seen as a liberating force associated with small-scale non-hierarchical development and change, there is another distinct facet to postmodernism. As well as the rejection of the modern and a hankering for the premodern, there is the establishment of 'after the modern'. This is frequently interpreted as 'consumerist postmodernism', involving the celebration of commercialism, commercial vulgarity, the glorification of consumption, and the related expression of the self (Cooke, 1990), trends which are of interest in relation to what is happening inside Third World cities (see Chapter 5). It is this aspect of postmodernism that gives rise to the suggestion that it maintains significant affinities with both conservative and radical lines of thought.

Such a condition is related to a conflation of trends in which aspects of art and life, high and low culture are fused together, or pastiched. Signs, hoardings and advertisements are potentially more important than reality. Mass communications lead to mass image creation (see Robins, 1989; Massey, 1991). History and heritage may be rewritten and reinterpreted in order to meet the needs of international business. This may all lead to further external control, exploitation and neo-colonialism.

In this regard, rather than being seen as a freeing, enabling and liberal force, postmodernism may alternatively be interpreted as the logical outcome of late capitalism (Harvey, 1989; Jameson, 1984; Dann and Potter, 1994; Potter and Dann, 1994; Sidaway, 1990; Cuthbert, 1995; Kaarsholm, 1995). The role of Transnational Corporations (TNCs) in the promotion of tourism in the Caribbean may be seen as an example of this, where advertising and promotion campaigns may be interpreted as being aimed directly at increasing both the environmental and social carrying capacities of the nation. Such developments have interesting implications in contexts where nations themselves are still endeavouring to modernise. Indeed the situation may give rise to all sorts of ambiguities (see Potter and Dann, 1994; Austin-Broos, 1995; Masselos, 1995; Thomas, 1991).

The development of postmodern trends which influence developing societies, in particular via the activities of TNCs and tourists, is of further interest in the new world order following the collapse of state socialism. We are certainly entering a noticeably less certain, less monolithic and unidirectional world. It seems highly likely that these trends will affect where social surplus product is concentrated, and in particular the degree to which it is urban-based (Potter and Unwin, 1995).

Conclusions

The present chapter has stressed that urbanisation is intimately associated with the generation and spatial concentration of social surplus product. Hence, it is tempting to argue that uneven development and inequalities appear to have been basic to urbanisation under capitalism throughout history. Together with Chapter 1, this chapter has also served to demonstrate that urbanisation cannot be understood in global isolation nor in an historical vacuum. Just as it is all too easy to omit that urbanisation and urbanism originated in what is currently referred to as the Third World, so it is tempting to succumb to the notion that contemporary Third World urbanisation represents an entirely different process. Whilst we might concur with Berry (1973) that there have been divergent paths in the urban experience of the twentieth century, equally we must recognise that present-day urban development in poor countries is a direct outcome of the past five centuries of global development and change. The processes that led to mercantile trade, distant colonialism and imperialism were also key agents responsible for promoting contemporary urban forms in developing countries. Thus, no account of contemporary Third World urbanisation can ignore the historical interdependence of the countries which now comprise the so-called First, Second and Third Worlds, however much the relations between these entities change in the new world order.

Chapter 3

National urban systems and global development

Brief perspectives on the nature of urban systems _____

As demonstrated in Chapter 2, much of development theory and practice can be interpreted as debating the role that towns and cities are seen as playing in the development process, and the precise balance that it is envisaged should be maintained between rural and urban areas. Chapter 2 considered this issue primarily from the point of view of theories and conceptualisations of development. The focus of the present chapter is on the main empirical generalisations that can be made about so-called urban systems. Chapter 4 then examines the policy implications of these characteristics. Hence, this chapter is quite strongly contemporary and futuristic in orientation.

The remit of the chapter is summarised by a simple framework for the study of urbanisation in developing societies, and this is shown in Figure 3.1 (overleaf). First and foremost, we are concerned with the nature and characteristics of urbanisation as they manifest themselves in particular territories. These include **urban primacy**, or the pre-eminence of one or more very large urban places over all other settlements. Primacy is frequently associated with differences between regions and the existence of **regional imbalances**. These forms of **uneven development** are often juxtaposed with **urban bias** and strong urban to rural flows – not just of people, but of surplus value. These conditions are also expressed in employment conditions, housing and the overall structure of towns and cities, topics which are dealt with in detail in Chapters 5 to 8. Secondly, the chapter is also concerned with the processes, both global and local, which are influencing urban systems and which are leading to such conditions. These include dependency and dependent relations (Drakakis-Smith, 1996a), and the impacts of increasing **globalisation** (see Anderson *et al.*, 1996; Allen and Hamnett, 1995; Allen and Massey, 1995; Massey and Jess, 1995). The latter process includes trends of **global convergence** and **divergence** in respect of consumption and production patterns respectively, which occur within the overall context of **postmodernity**. Such forces are also to be viewed in the context of recently executed **neo-liberal strategies** of development which, despite theoretical reorientations since the 1980s, are in practice promoting the salience of the market, via structural adjustment programmes, the deregulation of markets and principles of cost–recovery. These forces are all summarised in Figure 3.1 as overarching influences on the forms of urbanisation which are cutting across particular territories and regions.

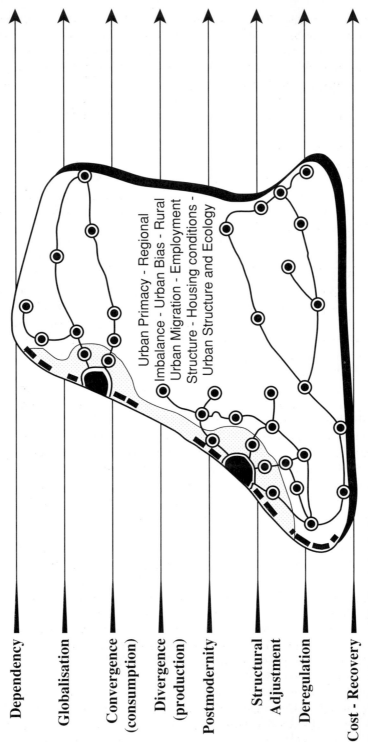

Dependency

Globalisation

Convergence
(consumption)

Divergence
(production)

Postmodernity

Structural
Adjustment

Deregulation

Cost - Recovery

Urban Primacy - Regional
Imbalance - Urban Bias - Rural
Urban Migration - Employment
Structure - Housing conditions -
Urban Structure and Ecology

Figure 3.1 A framework for the study of urban systems.

Finally, a simple working definition of an urban system is required. An **urban system** may be defined as the total set of towns and cities that together make up the settlement fabric of a given area, be it a region, nation or entire continental division. The urban system consists not only of a set of physical objects – in this case the urban settlements – but also of the flows and linkages which serve to bind them together. These flows are of people, capital, factors of production, ideas, information and innovations. By definition, urban systems are open systems which need to be studied in relation to their wider contextual environments. Urban systems are adaptive, continually changing and shifting their structures and external linkages. These properties of sets of towns and cities in area are, of course, well exemplified by the core–periphery and mercantile models of urban systems development, as detailed in Chapter 2. In particular, there is a need to examine urban systems from a distinctly **political-economy** point of view.

Uneven development and regional inequalities

It is clear from the accounts provided in Chapters 1 and 2 that urbanisation is frequently characterised by marked variations in economic development and welfare both across space and through time. Chapter 1 stressed that it is such differences between the urban and rural components of national space that are leading to rural-to-urban migration and ultimately contributing to the natural increase of population that is being recorded in the urban areas of developing countries. It was also implied that not only are urban areas as a whole generally becoming more prosperous in relative terms, but large cities are often becoming more affluent than small and intermediate urban places.

All of this may be viewed as part of a wider pattern of marked income divergence, regionally, personally and at the intra-urban level in developing countries. For example, the top 10 per cent of households in the Philippines account for almost 40 per cent of the total income and there is an almost non-existent middle class. Similar income shares for the top 10 per cent of households are recorded for India (33.6 per cent), Thailand (34.1 per cent), Malaysia (39.8 per cent), Venezuela (35.7 per cent), Mexico (40.6 per cent), Brazil (50.6 per cent), Zambia (46.4 per cent), Kenya (45.8 per cent) and Ivory Coast (43.7 per cent), as shown by the World Bank (1990) (see also Gilbert and Gugler, 1992). Taylor and Williams (1982: 9) stress how these personal income disparities in Third World cities exist alongside urban–rural disparities in wealth, so that a 'four-layer pyramidal structure occurs'. Thus, generally, a small wealthy urban elite exists at the apex, above secondly, government servants and those employed in modern industries, and thirdly, the vast numbers of the urban poor. Finally, the mass of rural poor exist at the broad base of the welfare pyramid.

The evidence shows quite clearly that **regional inequalities** in developing countries are substantially larger than those in developed countries. The data displayed in Table 3.1 give some impression of this. If the gross regional products of the richest and poorest regions of nations are compared, then

Table 3.1   Regional inequalities for a selection of developed and less developed countries

| Country | Gross regional product per capita, US$, 1976 | | Ratio of richest to poorest region |
	Richest region	Poorest region	
Developed areas:			
Belgium	4,380	2,616	1.67
France	5,918	2,833	2.09
Germany	7,022	2,683	2.62
United Kingdom	3,667	2,566	1.43
Netherlands	4,032	2,578	1.56
Italy	3,384	1,538	2.20
Japan	5,555	1,900	2.92
Less developed areas:			
India	217	97	2.24
Korea	582	270	2.16
Thailand	1,358	215	6.34
Malaysia	730	202	3.62
Iran	3,132	313	10.07
Colombia	1,342	199	6.75
Mexico	1,067	198	5.39
Brazil	1,102	109	10.14
Venezuela	1,354	237	5.72
Argentina	3,706	397	9.33

Source: Renaud, 1981, p. 118

typically developed countries show a ratio somewhere between 1.5 and 2.95. France, the Netherlands and the United Kingdom seem typical, with the richest region being up to twice as wealthy as the poorest. However, if we turn to developing nations, ratios up to 9 or 10 are not uncommon. Thus, in the case of Brazil, the ratio is over 10, and for Colombia, Mexico, Venezuela and Argentina it is above 5 (Table 3.1). Data presented by Gilbert and Gugler (1992) for the early 1980s place Brazil at 8, the Philippines at 6 and Peru at 14.

The vital issue is thus whether the development which has occurred will sooner, or even later, be accompanied by trickling-down or spread effects. However, there is much empirical evidence suggesting that many developing countries seem at present to be following the prescriptions of Myrdal rather than Hirschman. Thus, many commentators maintain that inter-regional disparities are showing little or no sign of reduction with development and change in developing countries. In Nigeria, for example, as noted in Chapter 1, some 76 per cent of manufacturing employment is concentrated along the southern coastal belt of towns. Similarly, in the case of Thailand, the Bangkok capital region accounted for 54 per cent of manufacturing value added, in relation to housing 10 per cent of the population (Gilbert and Gugler, 1992). In

Venezuela, it is estimated that rural incomes are on average only 40 per cent of those earned in urban areas. Similarly, in Barbados, wages in the agricultural sector average only 42 per cent of those in the non-agricultural sector (Hope and Ruefli, 1981; Potter, 1986b).

As discussed in Chapter 2, the classical view is that growth will slowly trickle down, and regions will become more equal over time. Williamson's (1965) frequently quoted empirical research based on a selection of European countries argued that the pattern of income inequality over time follows an inverted U-shaped trend. Thus, inequality increases during the early stages of development, but decreases thereafter. On this basis, Williamson maintained that Myrdal was excessively pessimistic, and he supported strongly the view of Hirschman that regional equality and income convergence would ultimately come about in an unplanned and spontaneous manner. It has been posited by neo-classicists that similar bell-shaped curves characterise changes in personal income inequalities within society (Kuznets, 1955), geographical concentration and urbanisation, demographic change and indeed development stages themselves over time, à la Rostow (Alonso, 1980). A number of writers have broadly concurred with Williamson's views, Mera (1973, 1975, 1978) and Alonso (1968, 1971) among them.

But for many present-day developing countries, the actual record seems to be one of marked and consistent inequalities. Gilbert and Goodman (1976) maintained that for every developing nation showing a tendency toward regional income equalisation, there is another that shows disequalisation. Similar arguments were expressed by Stöhr and Taylor (1981) and Friedmann and Weaver (1979). More recently, Gilbert and Gugler (1992) have commented that the limited data available suggest that whilst Brazil, Korea, Mexico and Peru have shown reductions in equality in the period since 1970, India, Indonesia and Thailand have displayed increases, whilst Colombia, Malaysia and the Philippines have experienced little or no change. Even in respect of the fast-growing and successful economies of Asia, Lo and Salih (1981) have shown that India, Malaysia, Thailand, the Philippines, Indonesia and Bangladesh all exhibited increasing regional income inequality over the latest period for which data were available, with only Taiwan, Sri Lanka and Pakistan recording trends towards equalisation (Figure 3.2, overleaf).

Friedmann and Weaver (1979) observe that regional planners appear generally to have hoped that the turning point for regional income equalisation is not far off, perhaps requiring only greater labour mobility, agricultural transformation and the transnational movement of capital to facilitate change from income divergence to convergence. However, a major point is that perceptions and stereotypes of inequality may well have a more pervasive effect on the perpetuation of regional inequalities than conditions themselves (von Böventer, 1975; Clarke, 1982), a suggestion that has strong links with Myrdal's theory of cumulative causation.

Enough Third World countries appear to be characterised by spatial polarisation and backwash effects to suggest that disequilibrium cannot merely be regarded as transitional and short-lived (Potter, 1986b). The issue effectively

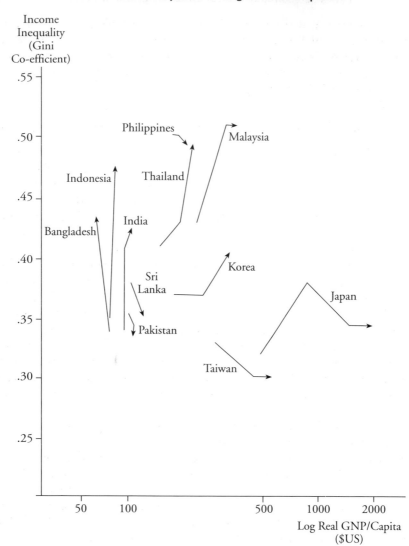

Figure 3.2 Income inequality against Gross Domestic Product for a selection of Asian countries (adapted from Lo and Salih, 1981).

boils down to the apparent conflicts existing between economic efficiency on the one hand and social equity on the other. Accordingly, the precise balance envisaged between these two goals stands as a crucial political and economic issue in development planning.

But there is a moral and ethical set of issues involved too. Even if equalisation is likely to occur in developing countries that are experiencing economic growth, how long must the poorest citizens wait before they too share in the fruits of growth and change? This issue is particularly salient, for Gilbert and Goodman (1976) argue that, for a number of reasons, regional income convergence is likely to be particularly weak in developing countries. They argue that one of these reasons is that regional differences in developing countries are now wider than those which were to be found in developed countries in the past; and another reason is the fact that developing countries may never reach the per capita income levels at which regional income equalisation occurs spontaneously, if indeed this ever happens.

Issues surrounding the condition of urban primacy

In recognising the scale of the regional inequalities that occur in developing countries, commentators are frequently drawn to discuss a particular type of spatial disparity which is referred to as urban primacy. Indeed, it is customary to assert that developing countries typically show such a condition.

Urban primacy was formally recognised by Jefferson (1939), and denotes a condition where the largest city in a country is superordinate in both size and national influence (see Figure 3.3A, overleaf). This contrasts with what is referred to as the **rank-size** or **lognormal** city size distribution, which is seen by some as being more typical of advanced industrial nations. The notion of the rank-size rule is generally attributed to Auerbach (1913) and Zipf (1949). It describes the situation where urban places follow a smooth progression by size (Figures 3.3B and C, overleaf). It is notable that in both the mercantile and core–periphery models reviewed in the last chapter, the early phase of economic development is characterised by increasing primacy. Further, in the full sequence of the core–periphery model, development and change are accompanied by the transition from a primate city size distribution to a smooth, or rank-size one.

However, even the most cursory evaluation of data shows that there is no simple relationship between type of city size distribution and the level of development of a country. Whilst it is the case that a good many developing countries exhibit urban primacy, there are also developed nations that do, France, Austria and Denmark among them. Equally, it is well known that some developing countries show low levels of primacy, at least as measured in relation to the upper echelons of their urban size distributions. Brazil, India and Kenya are frequently cited as particularly good examples.

Undoubtedly, the most important early paper dealing with this issue was published by Brian Berry (1961). The author looked at a sample of 38 countries, 13 of which showed clear lognormality, 15 strong primacy, and 10 distributions classed as intermediate between these two types. This research concluded that first, there was no clear statistical relationship between a country's city size distribution and its level of urbanisation. Secondly, the work indicated that different city size distributions were in no way related to the relative economic development of countries. Thereby, Berry suggested that at the national scale, no single variable influences city size distribution. Rather, he posited that a whole complex of forces influence relative city size distribution. In particular, it was suggested that if a few strong forces operate, urban primacy is the likely outcome. Berry observed that fewer forces are likely to influence the urban condition the smaller the country, the shorter its history

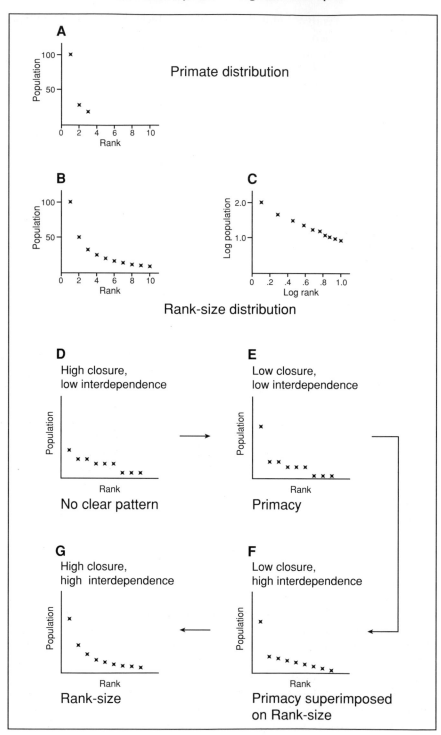

A

Primate distribution

B

C

Rank-size distribution

D

High closure,
low interdependence

No clear pattern

E

Low closure,
low interdependence

Primacy

G

High closure,
high interdependence

Rank-size

F

Low closure,
high interdependence

Primacy superimposed
on Rank-size

Figure 3.3 Various city size distributions.

Table 3.2   The relationships between urban primacy and a selection of other variables

Variable	Expected relation with primacy	Correlation
A. Linsky (1965):		
Areal extent of dense population	Negative	−0.37*
Per capita income	Negative	−0.22
Export orientation	Positive	+0.22
Colonial history	?	+0.21
Number in agriculture	Positive	+0.14
Rate of population growth	Positive	+0.33*
B. Mehta (1964):		
Gross National Product	Negative	−0.08
Level of urbanisation	Negative	−0.12
Overall population density	Negative	+0.02
Export dependency on raw materials	Positive	+0.19
Area of country	Negative	−0.28
Size of population of country	Negative	−0.29

* Statistically significant

of urbanisation, the simpler its economic and political life, and the lower its overall degree of development. Where the reverse conditions pertain, it is envisaged that a country will develop a range of specialised cities which perform a variety of functions, and a rank-size distribution will result.

These inferred relationships between city size distributions and wider social, economic and political variables were taken up by Arnold Linsky (1965), of the University of Washington School of Medicine. Linsky started his paper by pre-specifying the direction of the predicted relationships between urban primacy and six variables, as summarised in Table 3.2. Linsky suggested that primacy was positively associated with the degree of export-orientation of the nation, the proportion of the work force employed in agriculture, and the overall rate of population growth. The areal extent of dense population and per capita income levels were seen as being negatively correlated with primacy. Finally, an open verdict was initially pronounced on the association between primacy and former colonial status.

The results of cross-classifying urban primacy with each of the six variables for the 39 countries of the world which at that time contained at least one metropolitan area with over one million population, are also shown summarised in Table 3.2. All of the hypothesised associations between the variables and urban primacy were as expected. In addition, former colonial status was shown to be positively related to levels of primacy. The strongest relationship was a negative association between urban primacy and the size of countries ($q = −0.37$). The other statistically significant relationship was the positive one existing between primacy and the overall rate of population growth ($q = +0.33$). Linsky's work was important in demonstrating that whilst primacy is particularly characteristic of small nations which have low per capita incomes, a high dependence on exports, a former colonial history, an agricultural economy and a fast rate of population growth, it is certainly not precluded elsewhere.

At almost exactly the same juncture, very similar conclusions were being reached by Mehta (1964), a demographer. His results are also summarised in Table 3.2. Firstly, Mehta's analysis revealed no significant relation between primacy and Gross National Product, level of urbanisation or overall population density. As in Linsky's analysis, Mehta found that the strongest association was a negative one between urban primacy and size of population (−0.29), followed by a negative correlation with the areal extent of a country (−0.28). Urban primacy was also shown to be positively related to a nation's degree of export dependency on raw materials (Table 3.2).

Such work had shown the importance of variables such as size of country, population growth and export orientation in respect of primacy. In 1969, Vapnarsky provided fresh perspectives on the temporal manifestations of urban primacy. His work was empirically based on the history of economic development and changing city size distribution shown in nineteenth- and twentieth-century Argentina. Vapnarsky took what he described as an 'ecological approach', and argued that primate and rank-size patterns are not to be seen as the extreme ends of a continuum. Rather than being mutually exclusive as implied by so many writers, Vapnarsky argued that the two conditions are produced by different sets of circumstances.

Vapnarsky saw primacy as being positively associated with the degree of closure of the economy, that is its dependence on overseas trade. With increasing closure, urban primacy is thought to reduce, other things being equal. This accords well with the latter stages of the core−periphery model. Rank-size or lognormality, on the other hand, is regarded as being affected by the level of interdependence existing in a country, that is the extent to which regions are interlinked. Thus, as internal interdependence increases, so lognormality is progressively approached.

If these two quasi-separate variables are cross-classified, four principal types of city size distribution are recognised (Figure 3.3D to G, page 58). Following these clockwise from D to G in Figure 3.3 provides an interesting parallel with the city size distributions envisaged at each stage of the core−periphery model. Vapnarsky reviewed historical changes to the urban system of Argentina which suggests the veracity of this sequence. High closure with low interdependence (D) is typical of isolated and very underdeveloped areas, and is associated with no clear city size pattern. If interdependence remains low, but closure diminishes as a function of increasing trade (E), strong and growing primacy is the likely outcome. If as a third stage in the sequence, low closure is accompanied by internal interdependence (F), the result is primacy at the upper level of the urban system, juxtaposed by lognormality further down. Finally, if closure now increases, this plus the existing internal interdependence leads to the classical rank-size urban distribution (see Figure 3.3G).

These findings suggest that other things being equal, primacy is associated with a history of foreign dependency and low levels of internal interdepend-

Plate 3.1 A view of Caracas, Venezuela, a classic primate city in South America (photo: Rob Potter).

ence. These are both characteristics of former colonial territories (Plate 3.1). But clearly, a range of other factors is related to urban primacy, most importantly country size. The latter has been emphasised by Johnston (1971) in the case of the development history of New Zealand, where increasing closure and interdependence have been associated in a seemingly paradoxical manner with increasing primacy. Johnston explains this by virtue of New Zealand's small size, which has meant that import-substitution industrialisation has given rise to single-plant industries located in Auckland. Further, Johnston notes that during the period of European settlement, strong regional primacy became the norm, reflecting the orientation of local economies to the United Kingdom. Thus, given a doctrine of unequal growth, primacy may well be a logical response, especially in small countries and/or open developing ones. El-Shakhs (1972) has, in fact, shown that if less developed countries are considered on their own as a group, then a positive relationship exists between levels of urban primacy and relative levels of economic development, with such nations showing increasing levels of urban concentration during the early stages of development.

Clearly, there is a whole complex of forces which interact to mould the relative size characteristics of cities at the national level. Johnston (1980) emphasises the crucial point, in that urban primacy is normally examined at the level of the nation state. Further, he notes that these political divisions may be of little or no real consequence at the time of rapid urban systems development. In particular, as the New Zealand case bears testimony, colonisation from

Europe through a series of gateways is likely to have led to strong regional, rather than national primacy. This is particularly relevant in the case of large countries. For example, India shows regional primacy in terms of Calcutta, Bombay, Madras and Delhi. The same is true of China and Brazil. Thus, strong urban primacy may exist and be maintained at a regional level, but when aggregated, may sum to a distribution akin to lognormality. This type of argument had been advanced from an entirely theoretical perspective in the early 1960s by Berry and Barnum (1962). The important point is that even in countries where the level of national primacy is low, production, incomes and welfare levels may be highly skewed and uneven at the regional scale. Quite simply, primacy must not be regarded as being synonymous with polarisation, but rather as one very important form of it. For example, only 15 per cent of the total population of Brazil lives in São Paulo (Renaud, 1981), but as Table 3.1 (page 54) shows, Brazil's richest region returns a per capita regional product which is over 10 times greater than its poorest region. Similarly, a primacy level of only 16 per cent applies to Malaysia, but its regional inequality ratio is 3.62. Thus, although primacy frequently occurs in developing countries, it is not always apparent at the national level. Rather, the historical process of polarised growth via a number of cores has led to strong **regional primacy** and inter-regional disparities in most territories.

This section has demonstrated that it is not the case that there is a single trend from urban primacy to a rank-size pattern at the national scale with increasing development and change. It may suit the proponents of *laissez-faire* development and modernisation to assume that there is, but the empirical evidence does not substantiate this claim. Primacy, however, frequently occurs at the regional scale, especially in relatively small nations. Thus, size of country must be regarded as a potentially very important issue when considering the appropriateness of different urban and regional strategies.

Global shifts: industrialisation, diffusion and world cities

However, notwithstanding these comments on urban and regional inequalities, as noted in Chapters 1 and 2, the pursuit of unequal development as a matter of policy came to affect the newly independent, formerly colonial territories. It was perhaps inevitable that in seeking to progress during the postcolonial era they should associate development with industrialisation. This was hardly surprising given that the conventional wisdoms of development economics stressed so cogently this very connection. For many Third World countries, decolonisation afforded political independence and promoted the desire for the economic autonomy to go with it. In the words of Friedmann and Weaver (1979: 91) such nations:

> . . . took it for granted that western industrialised countries were already developed, and that the cure for 'underdevelopment' was, accordingly, to become as much as possible like them. This seemed to suggest that the royal road to 'catching up' was through an accelerated process of urbanisation.

In the early phase, the trend towards industrialisation in developing countries was closely associated with the policy of **import substitution**. This represented an obvious means of increasing self-sufficiency as such nations had traditionally imported most of their manufactured goods requirements in return for their exports of primary products such as sugar, bananas, coffee, tea and cotton. During the era of import-substitution industrialisation, key industrial sectors for development were those which were relatively simple and where a substantial home market already existed, for example, food, drink, tobacco, clothing and textile production. By now most developing countries have followed this path towards import-substitution industrialisation, although as Dickenson *et al.* (1983, 1996) observe, few have progressed much beyond it. The development of heavy industries, such as steel, chemicals and petrochemicals, along the lines of the Soviet model, has not been possible for most Third World nations.

Such a policy, which, following Rostow's (1960) linear model of development, might seem attractive, requires a level of population and effective demand not normally present in most developing countries. Further, the competition from developed nations, along with capital and infrastructural shortages and problems of lumpy investment, technological transfer and capital rather than labour intensity, also militate against such heavy industrial development. An exception, however, is provided by India which has achieved a high level of industrial self-sufficiency since 1945 (Johnson, 1983), and is now around the thirteenth industrial producer in the world.

However, from the 1960s, a number of developing countries embarked upon policies of light industrialisation by means of making available fiscal incentives to foreign companies. This policy of so-called '**industrialisation by invitation**' was strongly recommended by the Caribbean-born economist Sir Arthur Lewis (1950, 1955). It involved the establishment of branch plants by overseas firms, with the products being exported back to industrialised countries. The approach is closely associated with the setting up of Free Trade Zones (FTZs) and Export Processing Zones (EPZs). The FTZ is an area, usually located in or near to a major port, in which trade is unrestricted and free of all duties. The EPZ is normally associated with the provision of buildings and services, and amounts to a specialised industrial estate. Firms locating on them frequently pay no duties or taxes, and may well be exempt from labour and other aspects of government legislation. The approach is also frequently referred to as '*enclave industrialisation*'.

According to Hewitt, Johnson and Wield (1992) the first EPZ established in a developing country was at Kandla in India in 1965, and was quickly followed by further developments in Taiwan, the Philippines, Dominican Republic and on the United States–Mexican border. In 1971 nine countries had established such zones, and this increased to 25 by 1975 and 52 by 1985. In 1985 it was estimated that there were a total of 173 such zones around the world, which together employed 1.8 million workers. Frequently programmes of industrial development have been strongly urban-based, as in the case of

Table 3.3   Changes in the global distribution of industrial production, 1948–1984

Country/region	Percentage of world industrial production		
	1948	1966	1984
United States	44.4	35.2	28.4
United Kingdom	6.7	4.8	3.0
West Germany	4.6	8.1	5.8
France	5.4	5.3	4.4
Japan	1.6	5.3	8.2
Other developed countries	14.7	12.5	11.1
Centrally planned economies	8.4	16.7	25.4
Newly industrialised countries	4.9	5.7	8.5
Other less developed countries	9.1	6.5	5.4
	100.0	100.0	100.0

Source: Hewitt, Johnson and Wield, 1992, p. 18, based on Gordon, 1989

Barbados, where 10 industrial estates were established, all within the exisiting urban envelope (Potter, 1981; Clayton and Potter, 1996).

By such means, as summarised in Chapter 1, developing countries have increased their overall level of industrialisation. From 1938 to 1950, developing nations experienced a 3.5 per cent growth rate of manufacturing per annum, and from 1950 to 1970 this annual rate increased to 6.6 per cent (Dickenson *et al.*, 1983). But as stressed in Chapter 1, this growth has been minuscule by comparison with that of the urban population, and in many instances has been based on a non-existent prevailing level of industrial activity. Industrial growth has been characterised by two additional features. The first has been its highly unequal global distribution, and in the post-war period it has been associated with major changes in the global distribution of industrial production.

This process is illustrated in Table 3.3. The traditional industrial nations such as the United States and the United Kingdom, France, along with other developed countries, and latterly West Germany (before reunification), have all shown marked declines in their percentage share of world industrial production since 1948. This has gone hand in hand with rising industrial production in Japan, which by 1985 had increased its share of the world total to 8.2 per cent. Since 1948, industrial production has also risen sharply in the centrally planned economies. A major feature has been the increasing importance of the so-called **Newly Industrialised Countries (NICs)**, which by 1984 accounted for 8.5 per cent of global production.

However, the remaining less developed nations have shown a declining proportion of total manufacturing production, from 9.1 per cent in 1948 to 5.4 per cent in 1984 (Table 3.3). The emergence of NICs such as China, Brazil, India, South Korea, Mexico and Taiwan is shown by their inclusion among the top 25 industrial nations in 1986 (Table 3.4) (see also Dickenson, 1994; Courtney, 1994). The global distribution of manufacturing production

Table 3.4 The world's leading manufacturing nations in 1986

Rank	Country	Percentage of world total manufacturing value added
1	United States	24.0
2	Japan	13.7
3	USSR	12.2
4	China	10.5
5	West Germany	6.5
6	France	4.0
7	United Kingdom	3.5
8	East Germany	2.2
9	Italy	2.2
10	Canada	1.8
11	Brazil	1.7
12	Spain	1.1
13	India	0.9
14	South Korea	0.9
15	Mexico	0.8
16	Taiwan	0.8
17	Switzerland	0.8
18	Sweden	0.7
19	Netherlands	0.7
20	Romania	0.6
21	Poland	0.6
22	Czechoslovakia	0.6
23	Yugoslavia	0.6
24	Belgium	0.6
25	Argentina	0.6

Source: Dicken, 1992

is shown in Figure 3.4 (overleaf), and although the United States, western Europe and Japan between them account for three-quarters of total production, the importance of the Asian Tigers, together with Brazil and Mexico, is clear from the figure.

The process of change outlined above has been referred to by Dicken (1992) as **global shift**, whereby economic activity is becoming increasingly **internationalised/globalised**. This brings us to the second characteristic feature of post-1948 industrial change. This is the argument that **Transnational Corporations (TNCs)** represent the most important single force creating global shifts and changes in production (Dicken, 1992). TNCs can be traced back to the late nineteenth century, and early on dealt with agricultural, mining and extractive activities, but since 1950 they have become increasingly associated with manufacturing (Jenkins, 1987, 1992; Dicken, 1992; Sklair, 1994b; Daniels and Lever, 1996).

In 1985, the United Nations identified 600 TNCs operating in manufacturing and mining, each of which had annual sales in excess of $1 billion.

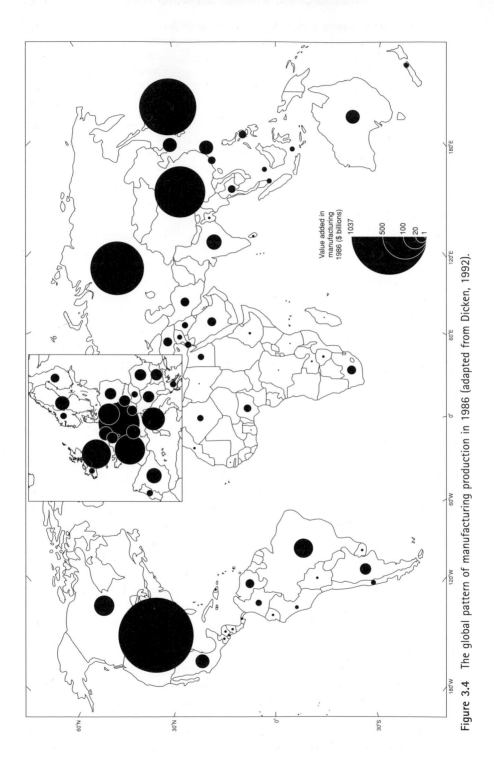

Value added in
manufacturing
1986 ($ billions)

1037
500
100
20
1

Figure 3.4   The global pattern of manufacturing production in 1986 (adapted from Dicken, 1992).

These corporations between them generated more than 20 per cent of the total production in the world's market economies. Clegg (1996) notes that multinationals originating in the 14 leading industrial countries numbered 7000 in 1969, but had increased to 24,000 by the early 1990s. Globally it is now estimated that there are around 37,000 multinational firms. Together these control around one-third of all private sector capital. During the 1960s, the foreign output of TNCs was growing twice as fast as world gross national product. By 1985, developing countries accounted for 25 per cent of total Foreign Direct Investment (FDI). The largest share of this was in Latin America and the Caribbean (12.6 per cent of the world total), followed by Asia (7.8 per cent) and Africa (3.5 per cent), with other areas accounting for 1 per cent (Dicken, 1992).

The fact is that these globalised patterns are far too complex – and contemporary – to be dealt with adequately by means of traditional theoretical and conceptual frameworks. In particular, the hierarchical model of change linked to modernisation which was put forward by Hudson (1969), Pedersen (1970) and Berry (1972), as reviewed in Chapter 2, would appear to be far too simplistic. Given the omnipotence of TNCs, it seems clear that new production, innovations, capital and social surplus are not trickling down the urban hierarchy in a step-by-step manner, from the top to the bottom. It can be argued that ownership and production are likely to be much more concentrated, an important theme which is picked up in the next section. Further, the decision to base production in one developing area rather than another will have considerable impact on the geography of development and change, especially when it is remembered that many TNCs have annual turnovers which greatly exceed the gross national products of many small and impoverished developing nations.

Just this sort of patterning has been identified by Pred (1973, 1977) in his historical examination of the growth and development of the urban system of the United States. Pred noted that the growth of the mercantile city was based on circular or cumulative causation, linked to multiplier effects. Further, Pred argued that the growth of large cities was based on their interdependence, so that large city stability has been characteristic. However, Pred maintained that key innovation adoption sequences were not always hierarchic, frequently flowing from a medium-sized city up the urban hierarchy, or from one large city to another. Pred (1977) looked at the headquarters of TNCs in post-war America, and stressed the close correspondence with the uppermost levels of the urban system. Thus, growth within the contemporary urban system is increasingly linked to the locational decisions of multinational firms and government organisations.

These types of notions have recently been given expression in the concept of the **world city**. Although nebulous in size-definitional terms, the idea is that certain cities dominate world affairs. At one level, this is a very straightforward and obvious proposition, but its contemporary relevance has been elaborated by Friedmann and Wolff (1982), Friedmann (1986) and Sassen (1991, 1996) under the title of **global city**. Friedmann (1986) put forward six hypotheses

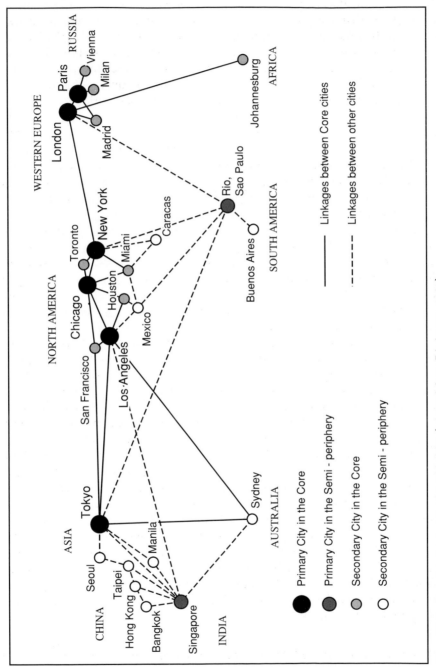

Figure 3.5 The global network of world cities (adapted from Friedmann, 1995).

about world cities, observing, for example, that they are used by global capital-ism as 'basing points' in the spatial organisation and articulation of production and markets, and that they act as centres for capital accumulation. Friedmann also suggested that the growth of world cities involves social costs which in fiscal terms the state finds it hard to meet. World cities have large populations, but more importantly they have large manufacturing bases, sophisticated finance and service complexes and act as transport and communication hubs, involv-ing TNCs and NGOs (Simon, 1992, 1993; see also Knox and Taylor, 1995; Friedmann, 1995; Sassen, 1991). Although the principal world cities such as New York, Brussels, Paris, London, Amsterdam and Milan are located in the developed world, Singapore, Hong Kong, Bangkok, Taipei, Manila, Shanghai, Seoul, Osaka, Mexico City, Rio de Janeiro, Buenos Aires and Johannesburg have all been recognised as part of an emerging network of world cities (Fried-mann, 1995). This emergence is given spatial expression in Figure 3.5. In short, world cities may be seen as points of articulation in a TNC-dominated cap-italist global system.

There is the implication, therefore, that uneven development is particularly likely to be associated with developing countries, and that their paths to devel-opment in the late twentieth century will be infinitely more difficult than those which faced developed countries. This argument has been reviewed in the case of poor countries by Lasuén (1973). He started from the premise that in the modern world, large cities are the principal adopters of innovations, so that natural growth poles become ever more associated with the upper levels of the urban system. Lasuén also observed that the spatial spread of innovations is generally likely to be slower in developing countries, due to the frequent existence of single plant industries, the generally poorer levels of infrastructural provision, and in some instances, the lack of political will. Thus, developing countries facing spatial inequalities have two policy alternatives. The first is to allow the major urban centres to adopt innovations before the previous ones have spread through the national settlement system. The second option is to attempt to hold and delay the adoption of further innovations at the top of the urban system, until the filtering down of previous growth-inducing changes has occurred. This may sound theoretical, but these options represent the two major strategies which can be pursued by states, and this theme is further illustrated in Chapter 4. The former policy will result in increasing economic dualism but, some would argue, the chance of a higher overall rate of eco-nomic growth. On the other hand, the latter option will lead to increasing regional equity, but potentially lower rates of national growth. As will be explored in Chapter 4, most developing countries have adopted policies close to the former alternative of unrestrained innovation adoption, seeking to max-imise growth rather than equity.

Global convergence and divergence and the urban system _____

This leads to a major conceptualisation of what is happening to the global system of cities in the contemporary world, and what this means for urban systems growth and structuring in present-day developing countries. The basic argument is that the uneven development that has characterised much of the Third World during the mercantile and early capitalist periods has been intensified post-1945, as a result of the operation of what may be referred to as the dual processes of **global convergence** and **global divergence**. The terms originate in the work of Armstrong and McGee (1985). Together these processes may be seen as characterising **globalisation** (Allen and Hamnett, 1995).

Divergence relates to the sphere of production and the observation that the places which make up the world system are becoming increasingly differentiated, that is diverse and heterogeneous. Starting from the observation that the 1970s witnessed a number of fundamental shifts in the global economic system, not least the slowing down of the major capitalist economies and rapidly escalating oil prices, Armstrong and McGee stress that such changes have had a notable effect on the urban systems of developing nations.

Foremost among these changes has been the dispersion of manufacturing industries to low-labour cost locations, and the increasing control of trade and investment by the TNCs. It is this trend which has witnessed the establishment of Fordist production-line systems in the NICs, whilst smaller-scale, more specialised and responsive, or so-called flexible systems of both production and accumulation have become more typical of western industrial nations. In this fashion, productive capacity is being channelled into a limited number of urban centres. Thus, increasing global division of labour and the increasing salience of TNCs are leading to enhanced heterogeneity or divergence between nations with respect to their patterns of production, capital accumulation and ownership. Thus, the industrialising export economies of Taiwan, Hong Kong and South Korea may be recognised, along with the larger, internally directed industrialised countries such as Mexico, raw-material exporting nations like Nigeria, and low-income agricultural exporters such as Bangladesh. As argued in the previous section, such changes are non-hierarchic in the sense that they are focusing development on specific cities and localities. So Armstrong and McGee (1985: 41) state that 'Cities are . . . the crucial elements in accumulation at all levels, . . . and the *locus operandi* for transnationals, local oligopoly capital and the modernising state'. It is these featues which gave rise to the title of their book, which described cities as 'theatres of accumulation'.

On the other hand, many commentators point to what ostensibly appears to be the reverse trend – that of the increasing similarity which appears to characterise world patterns of urbanisation. There is at least one respect in which a predominant pattern of what may be referred to as global **convergence** is occurring. This is in the sphere of consumer preferences and habits. Of particular importance is the so-called '**demonstration effect**', involving the rapid assimilation of North American and European tastes and consumption patterns (McElroy and Albuquerque, 1986).

The influence of the mass media, in particular television, videos, newspapers, magazines and various forms of associated advertising, is likely to be especially critical in this respect. The televising of North American soap operas may well lead to a mismatch between extant lifestyles and aspirations (Miller, 1992, 1994; Potter and Dann, 1996). Such media systems have become truly global in character in the 1990s (Robins, 1995). Potter and Dann (1996) show that the ownership of televisions and radio receivers is near-universal, even among low-income households, in Barbados in the eastern Caribbean. Further, a surprisingly high proportion of households there have a video recorder, some 43.24 per cent in 1990; video ownership was as high as 27.82 per cent for the occupants of all-wood houses, and 48.26 per cent for the residents of combined wood and concrete houses, those which are generally in the process of being upgraded. Other aspects of the wider trend of convergence involve changes in dietary preferences, and the rise of the '**industrial palate**', whereby an increasing proportion of food is consumed by non-producers (see MacLeod and McGee, 1990; Drakakis-Smith, 1990).

Developing cities may be seen as the prime channels for the introduction of such emulatory and imitative lifestyles, which are sustained by imports from overseas along with the internal activities of transnational corporations and their branch plants. In turn, these are frequently related to collective consumption, indebtedness and increasing social inequalities. Such changes toward homogenisation are ones which are particularly true of very large cities. However, it is the elite and upper-income urban groups who are most able to adopt and sustain the 'goods' thereby provided – for example, health care facilities, mass media and communications technologies, improvements in transport and the like. It may be conjectured that the lower income groups within society disproportionately receive the 'bads' – for example, formula baby milk and tobacco products. Thus, convergence is likely to have a differential social impact on Third World populations and will not lead to homogenisation *per se*. Indeed, there is a strong argument that such changes operate on, and further exploit, the differences which exist between local populations. The implications of the process of convergence at the level of the individual city will form one of the main themes of Chapter 6.

A direct and important outcome of this suggestion is a strong argument that the form of contemporary urbanisation found in particular areas of the developing world is the local manifestation of the juxtaposition of these two seemingly contradictory processes at the global scale. In terms of examples, Armstrong and McGee look at the ways in which these two processes are played out in Ecuador, Hong Kong and Malaysia. Potter (1993b, 1995) has examined how well the framework fits in the examination of Caribbean urbanisation where, it is argued, tourism has a direct effect in respect of trends of both convergence and divergence.

We are now at a point where a number of important arguments can be reconciled. The first is that it can be posited that it is the key traits of western consumption and demand that are being spread in an **hierarchical manner** within the global urban system, from the metropolitan centres of the core world cities to the regional primate cities of the peripheries and semi-peripheries, and then subsequently down and through the urban system. It is interesting to observe that the innovations that were cited by Berry and others in the 1960s

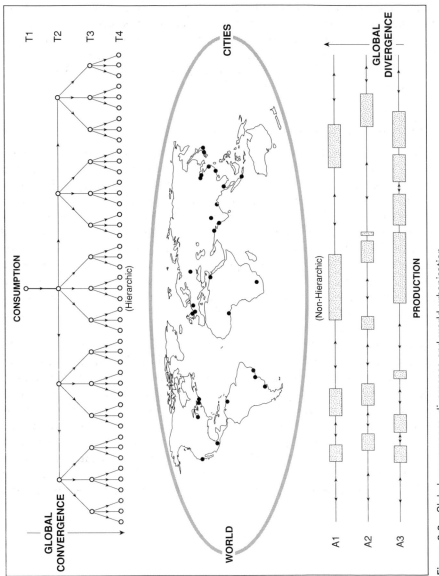

Figure 3.6   Global convergence–divergence and world urbanisation.

and 1970s as having spread sequentially from the top to the bottom of the urban system of America were all consumption-oriented, for example the diffusion of television receivers and stations.

In contrast, aspects of production and ownership are becoming more unevenly spread, being concentrated into **specific nodes**. This process involves strong cumulative feedback loops. Hence considerable stability is likely to be maintained at selected points within the global urban system, frequently the largest world cities. Hence, entrepreneurial innovations will be concentrated in space, and are not likely to be spread through the urban system, an argument which has parallels with the view that sees dependency theory as the diffusion of underdevelopment rather than development.

Conclusions

The key elements of the argument presented in the last section are summarised in Figure 3.6. On the one hand, cities serve to diffuse the culture and values of westernisation. By such means, patterns of consumption are spread through time (T1 . . . T2 . . . T3 etc.), and there is an evolving tendency for convergence on what may be described as the global norms of consumption. Figure 3.6 recognises that these aspects of global change are primarily expressed hierarchically, and are top-down. In contrast, cities appear to be concentrating and centralising the ownership of capital, and this process is closely associated with differences in productive capabilities. The tendency toward divergence is expressed in a punctiform, sporadic manner, which stresses activities in area (A1 . . . A2 . . A3). TNCs and industrialisation are the most important agents involved in this process.

The most important point is that cities and urban systems have to be studied as important functioning parts of the world economy. In such a role, cities are agents of concentration and spread at one and the same time. Viewed in this light, the age-old argument as to whether cities are generative or parasitic is seen as naive, simplistic and misfounded. It is also far too simple to ask if cities are getting more similar or not around the world (Potter, 1997). Similarly, it is far too simplistic to ask whether cities spread change in a hierarchical or non-hierarchical manner, for in fact, they are doing both simultaneously. In this regard, it is tempting to argue that the breaking down of rigid hierarchical systems at a global level is very much part of the **postmodern world**.

Chapter 4

National urban development strategies

Introduction: planning and the state

A suitable juncture has been reached at which to consider the practical policy and planning aspects of the debates which surround the issues of urban primacy, over-urbanisation, unequal development, urban bias, and regional inequalities in developing societies. Policies which endeavour to regulate and modify existing patterns of urban development at the national–regional scale are referred to as **national urban development strategies**. Naturally, specific national urban development strategies have a close relationship to one or more of the broad theories and conceptualisations of urbanisation which have been reviewed in Chapters 2 and 3. The division of policies into those which are **top-down** and those which are **bottom-up** is particularly relevant.

Notwithstanding the arguments and counter-arguments presented in Chapters 2 and 3, most countries have at some stage made public declarations of their intention to reduce congestion in the primate city or cities, or to reduce the manifest inequalities existing between the regions which make up the nation. Put another way, if a panel of experts were convened in order to ponder the problems of urbanisation in the developing world then, like as not, they would agree that planned urban decentralisation or deconcentration is required (Potter, 1995). However, as noted by Gilbert and Gugler (1992), despite frequent planning documents and statements of intent, few countries have earnestly endeavoured to control the growth of large cities, or to decentralise people, industry and other jobs from core regions, or to reduce rural poverty. And for those that have, a strong political bias exists, specifically towards the left.

It is entirely possible, of course, that the state may be involved in making one set of statements whilst it is doing something quite different in the policy arena. Specifically, the state may talk about promoting regional equality, but do little or nothing to actively promote it. It is in this sense that quite frequently, there are mismatches between planning on paper and planning in reality. Of course, such mismatches between rhetoric and policy may not always be as regressive as they at first sound. As an example, from 1970 onwards, the government of Barbados declared its intention of decentralising people and activities away from the primate capital, Bridgetown (see Potter and Hunte, 1979; Potter, 1981, 1983). In 1978, the United Nations conducted a survey of governmental attitudes to population distribution among

Latin American and Caribbean countries (see Peek and Standing, 1982; Table 1: p. 2). The results showed that of the 27 respondent countries, as many as 21 reported that they regarded their population distributions as extremely unacceptable, four that they were substantially unacceptable, and one, Cuba, that it was slightly unacceptable. Only one national government reported that it considered its existing population distribution to be entirely acceptable. That country was Barbados, despite its espousal elsewhere of the singular aim of decentralisation as its major policy plank since 1970.

This example stresses the importance of theories of the state in considering urban and regional policies. Whilst in the context of a parliamentary democracy the state may be seen as the supplier of public goods and services, a benign regulator and facilitator, a social engineer, or an arbiter of conflicting claims on resources (Dear and Clark, 1978; Clark and Dear, 1981), in other contexts the reality may be very different. In the 1970s, Miliband (1977) coined the phrase '**the state for itself**', to describe situations where the state's actions are basically designed to ensure the maintenance of the interests of the ruling group, and not those of the general populace. Thus, those in powerful positions may effectively use their power to further their own economic ends, along with those of their families, associates and friends. Others have employed the term '**bureaucratic authoritarianism**' where, for instance, oil-rich states such as Nigeria and Venezuela have used their revenues to provide themselves with inordinate power (see Gilbert and Healey, 1985). Sandbrook (1985) argues that '**neopatrimonialism**', or personal rule, has seriously impeded development in post-colonial Africa. In examining corruption in the Dominican Republic, Vendovato (1986) likewise made use of terms such as the '**predatory state**', and '**kleptocracy**', in which public office is used for the illegal acquisition of personal wealth. In such circumstances, as noted by Potter (1985), just as one person's loss is another's gain, so one region's loss is potentially another region's gain, so that inequality may be maintained or even enhanced, whilst at the very same juncture, political rhetoric runs against it (see also Potter and Binns, 1988).

Many writers in the field of urban and regional planning studies have argued that only socialist states such as Cuba have seriously endeavoured to reduce urban and rural imbalances in national and regional growth. It is possible to suggest a simple model of urbanisation under different levels of state intervention. This is based on the framework provided by Murray and Szelenyi (1984) (see also Forbes and Thrift, 1987; Potter, 1992b). This is shown in Figure 4.1 (overleaf). The diagram firstly suggests that urbanisation has tended to advance most quickly under free-market or capitalist conditions. The diagram also implies that rates of urbanisation have most frequently been reduced in countries where the state has become increasingly directly involved in the development process. Some socialist Third World countries have followed what amount to **anti-urban policies**, as a result of cities being seen as being associated with elite groups and foreign influences. Some countries, such as South Vietnam between 1975 and 1980, have followed a path of **deurbanisation**. A less drastic alternative is **zero urban growth**, where the urban status

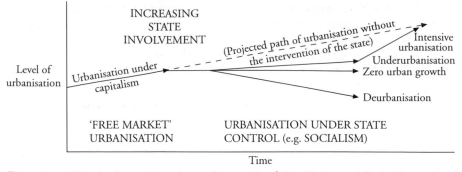

Figure 4.1   Urbanisation paths and state intervention (after Murray and Szelenyi, 1984; Forbes and Thrift, 1987).

quo is maintained; China basically followed such an approach during the period 1958–75. In other instances, the state may promote what is referred to as **under-urbanisation**. This occurs where urbanisation proceeds, but at a rate which is slower than the growth of employment in industry (see also Simon, 1990).

But Figure 4.1 also acknowledges that urbanisation does not always occur slowly under state intervention, and in some socialist countries, and indeed state-controlled mixed economies, **intensive urbanisation** is promoted. This might occur, for example, after a period of deurbanisation, zero urban growth or under-urbanisation.

Questions of city size: restrained versus unrestrained urban growth

For some time during the 1960s and 1970s, the literature was concerned with investigating whether there was an ideal or optimum city size. If cities are able to reap the benefits of economies of scale, then in theory there should be a population size at which these advantages are maximised. Economies of scale include those which are both internal and external to the firm, along with pure economies of agglomeration. In the end, however, this approach turned out to be just what it purported to be – an entirely theoretical exercise. Thus, the search for an optimal city size was eventually described by one of its principal proponents as a game – 'fun even though it gets nowhere' (Richardson, 1978: 322).

One of the most obvious criticisms of the optimal city size approach is that efficiency is required at a whole series of different size levels, so that by definition there cannot be a single optimum city size. The same argument can be extended to suggest that city size will vary according to socio-cultural context. In a parallel argument, some have questioned whether the growth of large cities should be curtailed in any way. In the 1970s, in particular, a number of economists and regional scientists followed the neo-classical approach, arguing that large metropolitan centres always produce more benefits than costs, and that any attempt to retard the growth of the largest cities would be likely to reduce national rates of economic growth (Richardson, 1973b, 1976; Alonso, 1971; Hoch, 1972).

However, in a series of exchanges with Richardson, Gilbert (1976, 1977) took issue with the doctrine of unrestrained urban growth, especially in developing countries. From a primarily economic standpoint initially, Gilbert (1976) questioned the assumption that higher productivity in big cities is brought about by economies of scale. Rather, he suggests that this is achieved at the expense of productivity elsewhere. It was suggested that if 'infrastructure of the same quality were provided in medium-size centres, then productivity in these cities would rise' (Gilbert, 1976: 29). This is, of course, tantamount to saying that capitalist development has promoted a circular argument. It has promoted the productivity of large cities by concentrating investment in them, and has subsequently claimed that only large cities are productive. However, as a regional economist who has consistently argued the case for the productivity of large cities, by the end of the 1980s Richardson (1989) had admitted that the productivity arguments that he had advanced for large cities for some years might well have been overstated.

The precise veracity of these arguments is difficult to establish, and there are relatively few data available on this topic. Gwynne (1978) shows that the average index of retail prices for Chilean cities exhibits an inverted U-shaped pattern with increasing city size. Thus average prices were low in the primate capital Santiago, and also for small towns in the 20,000–60,000 population bracket. On the other hand, prices were on average higher in the intermediate size places with populations of between 60,000 and 250,000. Gwynne concludes that these places are too small to benefit from the economies of scale which are afforded by large marketing organisations, and too large to promote efficiency in traditional marketing systems. This example serves to re-emphasise that the efficiency of cities is not likely to be linked to city size in some simple manner, and that efficiency is required at all levels of the system.

Perhaps the most salient point is that arguments concerning large cities should not be based on economic reasoning alone. Quite simply the issue depends on the goals that are set, and these, of course, are socio-political, cultural and ultimately moral–ethical. This in turn stresses the balance that is promoted between social equity and economic efficiency.

National urban development strategies

Overview

Despite such conflicting views concerning the role of very large cities, attempts to decentralise people, jobs and social infrastructure away from primate cities and congested metropolitan regions can be seen as the single most frequently pursued urban planning objective in both developed and developing nations alike. As suggested above, these policies may frequently reflect social, political and strategic motives rather than economic ones. However, there is all too

CONCENTRATED URBANISATION

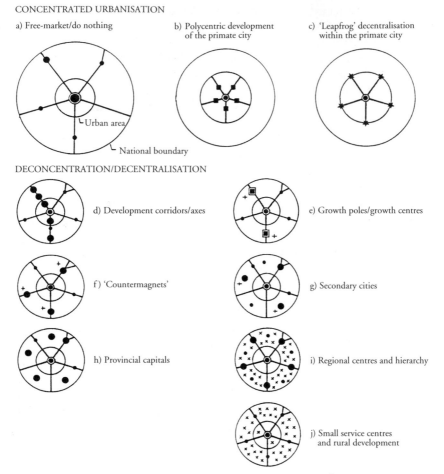

a) Free-market/do nothing

b) Polycentric development
   of the primate city

c) 'Leapfrog' decentralisation
   within the primate city

DECONCENTRATION/DECENTRALISATION

d) Development corridors/axes

e) Growth poles/growth centres

f) 'Countermagnets'

g) Secondary cities

h) Provincial capitals

i) Regional centres and hierarchy

j) Small service centres
   and rural development

Figure 4.2   A selection of national urban development strategies (based on Richardson, 1981).

frequently the suspicion that the size of urban areas is being treated as if it were itself the problem, rather than one manifestation of a series of wider socio-economic and political problems.

Richardson (1981), although as noted earlier having previously argued from a strongly pro-large city standpoint, reviewed the importance of national urban development strategies in developing countries. In his paper, 10 prototype strategies representing 'attempts to change the inter-urban distribution of population in the pursuit of some long-term policy objectives' were identified (Richardson, 1981: 267).

These have been given a spatial representation by the present author in Figure 4.2. Given an existing large city at the centre of a hypothetical circular national state, there are three policies which can be regarded as enhancing concentrated urbanisation. A hypothetical primate city is shown at the centre of a circular national territory in Figure 4.2a. The primate city region is shown by a dotted circle, and the other important urban areas beyond it by five black circles. The first policy of concentrated urbanisation is the *laissez-faire*, free-market or do nothing one of letting the market take its course. If, however, problems of congestion and diseconomies are recognised in the primate city, efforts may be made to decentralise, but merely within the core region. This may be regarded as an intra-urban response to what is essentially an inter-regional problem. Thus, a polycentric pattern of settlement may be promoted in the primate city (Figure 4.2b), or a form of leapfrog decentralisation to the edge of the core region (Figure 4.2c) may be envisaged.

Strategies (d) to (j) in Figure 4.2, on the other hand, all represent various forms of dispersion at the inter-regional scale, and involve spatial deconcentration–decentralisation from the primate core region. In strategy (d), development corridors/axes, growth is focused along one or more principal inter-urban corridors, whilst in strategy (e) growth is channelled into one or more dynamic growth poles or centres, the effectiveness of which we shall examine later in this chapter. A variation on the same theme is the strengthening of a few major distant nodes as countermagnets, as shown in Figure 4.2f. Other forms of decentralisation can be effected by the promotion of a limited number of secondary cities (strategy (g)), or by the establishment of provincial state and departmental capitals (strategy (h)). Strategy (i) involves the promotion of regional metropolises and an associated hierarchy of urban places. At the far end of the scale, a dispersed policy of small service centres and associated rural development throughout the periphery may be pursued (j). These strategies are not mutually exclusive and indeed several of them are very similar indeed and can be combined into hybrid policies.

Richardson is at pains to stress that the reasons for the implementation of such policies may be social rather than economic, so that the 'key goals of a National Urban Development Strategy are the same as societal goals in general' (Richardson, 1981: 270). Further, urban development strategies have to be highly country-specific if major blunders are to be avoided. For instance, the size of a country is a particularly crucial dimension, as shown in Chapter 3 in relation to city size distributions. Richardson finishes by continuing to argue that the efficiency merits for slowing down urban primacy are dubious; but it must be stressed that he is arguing from a basically economic, growth-oriented viewpoint, rather than a wider socio-economic position.

Growth poles and growth centres

The main thrust of national and regional urban planning policies in many countries in the 1960s was the strategy of 'concentrated deconcentration', whereby strong polarisation was replaced by promoting polarisation elsewhere. Generally speaking, this approach was mediated through the idea of growth poles and growth centres. As explained in Chapter 2, the original idea of the growth pole lay in the works of Schumpeter (1911) and Perroux (1950, 1955). It initially referred to a set of expanding industries located in an urban area, which would thereby induce further innovations, linkages and economic development.

It will be recalled that historically Myrdal, Hirschman, Friedmann and Vance had all agreed that the early stages of economic development have normally been associated with spatial concentration in a single core region. This, of course, is associated with the idea of the 'natural' or 'spontaneous' growth pole. On the other hand, Myrdal and later Friedmann argued the need for intervention if development and change are ever to be spread. The notion of planned polarisation reversal gives rise to the planned growth pole. As Brookfield (1975) noted, however, the original sectoral idea of the growth pole was progressively widened eventually to become synonymous with an urban centre containing expanding economic activities. This geographical development of the concept owed much to the work of Boudeville (1966). The approach really represented the evolution of the spatially conceived growth centre, as opposed to the sectorally conceived growth pole (Richardson, 1978; Brookfield, 1975).

Growth centre and growth pole policies have been the subject of considerable academic debate and many detailed expositions appeared during the 1960s and 1970s (see Hansen, 1967, 1972; Darwent, 1969; Lasuén, 1969; Kuklinski, 1972; Parr, 1973; Moseley, 1974; Buttler, 1975). The employment of such techniques for practical planning purposes has been just as frequent, first in European countries, but later in developing countries as well, where they seemed to dovetail with the idea of reducing urban primacy. As noted by Richardson (1978), in such contexts the policy can serve to link national economic policy to intra-regional strategies by filling the gaps existing in the urban system. Examples abound: for instance, regional physical development plans in Kenya have incorporated the growth pole concept, as they have also done in Tanzania, India, and the socialist states of China and Poland. Growth pole policies have also been employed frequently in Latin American countries, as shown in the review by Richardson and Richardson (1975). For instance, in Chile in the late 1960s, a growth pole was established in each of the 11 main regions of the country. In the case of Venezuela a planned growth pole based on heavy industry was developed at Ciudad Guyana and a secondary pole at Ciudad Bolivar, also in the east of the country (see Turner and Smulian, 1971). Similarly, in Colombia, after 1969, the primacy of Bogotá was planned for by means of the expansion of the next three urban places of Medellin, Cali and Barranquilla.

The growth pole concept has been applied in a large number of different ways and by promoting poles of widely different size. As a result, some analysts have argued that it has become a very diffuse and weak idea (Moseley, 1973; Gilbert and Gugler, 1992). The somewhat cloudy reputation of the growth pole is exacerbated by the suggestion that, in practice, it has failed to live up to its promise (Richardson, 1978). This partly reflects the problem of identifying an appropriate population base for growth poles, but is also a reflection of the very long time horizon that appears to be needed for the establishment of a growth pole or centre.

More fundamentally, it can be argued that the whole concept is basically Western, in both origin and philosophy. From a radical stance, it is argued that growth pole policy is based on the dispersion of polarisation and urban bias at the national and regional scales and not its ultimate eradication. Hence, from a Marxist perspective, it is ventured that growth poles are dependent on principles of unequal development and the continuing dominance of western nations. However, despite these criticisms, some have argued that the policy is not without its potential merits, especially if a link is forged between growth pole policies and rural-based strategies of development in developing countries.

The missing middle of the urban system in developing countries

A theme which runs through all of the national urban development strategies initially reviewed is the underdevelopment of the middle tiers of the settlement hierarchy. This line of argument was strongly followed in the Third World context in a book published by Johnson (1970), an economic historian. In his work on spatial organisation in developing countries, and in his earlier volume based on India (Johnson, 1965), Johnson was taken with the general paucity of towns in developing regions. For example, Johnson (1970) showed that the nations of Europe had approximately 10 times as many central places (towns over 2500 people) per village, as the Middle Eastern. It was suggested that the gap in the national settlement system must serve to lower agricultural productivity in developing countries. This led Johnson to argue strongly for town-building programmes as a major plank of economic development policy in developing countries.

Johnson's work was clearly influenced by the classical central place theories of Christaller (1933) and Lösch (1940). As an example, the planning of the national settlement system of Israel after 1948 followed some elements of this type of approach. Thus, in 1948, 43 per cent of the total population lived in Tel Aviv. In the next 20 years, some 34 new towns were created, accounting for 21.3 per cent of the national population. As a result, by 1970, Tel Aviv's share of the national population had fallen to 33 per cent, and a tendency to rank-normality was discernible.

In assessing Johnson's overall thesis, some authors have argued that in reality the causation is the reverse of that which he suggests, and that the lack of urban places reflects the low level of agricultural production (Gilbert and Gugler, 1992). Certainly, it is hard to see how changes such as those suggested by Johnson would achieve the desired end in the absence of more radical reforms.

Despite such caveats, the contemporary policy-relevance of the argument that the middle orders of the urban system need to be bolstered was emphasised by an American planner-cum-political scientist, Dennis Rondinelli, in the 1980s (Rondinelli, 1982, 1983a, 1983b; see also Cheema and Rondinelli, 1983), and picked up by others (Hardoy and Satterthwaite, 1986). Rondinelli started from the observation that since the 1970s, an increasing emphasis has been placed on rural development as a corrective to the earlier urban bias in foreign aid and development programmes. With this it is argued that a policy emphasis on developing middle-sized cities has also emerged in many countries, in order to reverse, or at least reduce, the worst effects of polarisation, and to commercialise agriculture in rural regions. These ideas are closely linked to those expressed by Johnson, although Rondinelli envisages such 'intermediate' or 'secondary' cities as places with over 100,000 population, but which are considerably smaller than the largest city in the country. In particular, Rondinelli (1982) maintains that urban places of this size can be used in order to promote a more equitable distribution of population and resources than can be achieved by establishing a few selected growth areas.

Bottom-up approaches to settlement planning

Following the arguments presented in Chapters 2 and 3, it should come as no surprise that a growing number of regional planners and development theorists have since the late 1970s suggested that the growth centre policy is deficient in that it is directly premised on the principles of development from above. The concept of rural development associated with more fundamental changes is connected with the doctrine of development from below, as reviewed in Chapters 2 and 3. This is particularly associated with the work of Friedmann and Douglas (1975) and, more recently, Friedmann and Weaver (1979), and is referred to as 'agropolitan development', or 'urban-based rural development', as noted in Chapter 2.

Such an approach can be related to Michael Lipton's (1977) thesis that Third World poverty is largely the product of urban bias in world development imperatives, so that the 'poor stay poor'. Although Lipton associated urban with high status and rural with low status in a rather over-generalised manner, the idea that in development terms urban areas have been accorded far too many advantages is a very tempting one.

The starting point for the agropolitan philosophy is that polarised development and the growth pole theory have acted as the 'handmaiden(s) of transnational capital' (Friedmann and Weaver, 1979: 186) and have been 'completely attuned to transnational corporations' (p. 188). Under such conditions, it is argued that development has been dominated by purely functional principles. In contrast, Friedmann and Weaver urge for a return to a **territorial basis** for global development, founded on 'basic needs'. This partly reflects Johnson's (1970) thesis, but more significantly reflects a move away from the 'fetishism of growth efficiency' (Friedmann and Weaver, 1979: 195).

As noted in Chapter 2, a principal idea is that basic needs must first be satisfied within areas and this is associated with selective regional–territorial closure in economic terms, and also with the communalisation of productive wealth. It is envisaged that later, the economy can be diversified and non-agricultural activities introduced. The suggestion is made that when industrial activities are brought in, an urban location is not mandatory and that hitherto agglomeration economies have been greatly overstated. Thereby, it is envisaged that agropolitan development will lead to cities that are based on agriculture. The approach as a whole is designed to build on strength from within and is

Table 4.1   Stöhr's eleven essential components of development from below

1.	Broad access to land
2.	A territorially organised structure for equitable communal decision-making
3.	Granting greater self-determination to rural areas
4.	Selecting regionally appropriate technology
5.	Giving priority to projects which serve basic needs
6.	Introduction to national price policies
7.	External resources only used where peripheral ones are inadequate
8.	The development of productive activities exceeding regional demands
9.	Restructuring urban and transport systems to include all internal regions
10.	Improvement of rural-to-urban and village communications
11.	Egalitarian societal structures and collective consciousness

Source: based on Stöhr, 1981

premised on the skills and resources of the local population, and not external aid and capital (see Table 4.1). However, judicious use of the latter is not ruled out later.

Such an approach, involving the recovery of the territorial bases of life, seems very attractive given the overwhelming problems of urban-based development currently being experienced in Third World countries, not least in environmental terms (see Chapter 9). It is argued by Friedmann and Weaver (1979: 200) that eventually 'large cities will lose their present overwhelming advantage'. It is clear that the agropolitan approach has political connotations and is linked in the main, but not exclusively, to socialist policies. Prime examples have included China, Vietnam, the Republic of Korea, Sri Lanka, Bangladesh, Pakistan and Tanzania. The involvement of the masses of the people by means of active public participation at the local level becomes a vital prerequisite of the agropolitan approach.

One of the productive aspects of the approach is that it suggests the relevance of the question 'why does growth have to be based on large cities?'. The countervailing argument that development planning might just as easily be based on medium-sized, or indeed small, urban places in a predominantly rural context, comes into focus in the policy arena as the outcome of such deliberations (Lowder, 1991; Drakakis-Smith, 1996b).

Examples of national urban development strategies _____

As discussed earlier in this chapter and in previous ones, large cities have frequently been bolstered within capitalist nations since the 1940s. When in such contexts the economic virtues of large cities have come into question, some form of deconcentration has been attempted, but all too often by promoting planned agglomeration elsewhere, by means of employing growth poles and the like. Hence, several writers have asserted that outside socialist countries, few if any real attempts have been made to control the growth of metropolitan areas and to reduce spatial inequalities. It is the socialist countries such as Cuba, China, Cambodia and Tanzania that have made a more concerted effort to control or reduce the growth of large cities. As Stretton (1978) explains, communist governments have generally been suspicious of city living and uncertain as to how they should respond (see Figure 4.1, page 76). Such states have frequently moved towards anti-urban policies, or at the very least, strongly rural-based strategies of development (Stretton, 1978). It is this which makes them worthy of consideration as examples here.

China

China is a model of the agropolitan approach advocated by Friedmann. Since 1949, predominantly anti-urban and pro-rural policies have been adopted (see Schenk, 1974; Wu and Ip, 1981; Ma, 1976; Kirkby, 1985; Ma and Noble, 1986; Pannell, 1990; Tang, 1993). However, as Chang (1982) notes, policies have changed periodically since 1949. At some points strong arguments have been advanced for deurbanisation. During others, the importance of centralised planning and major urban-based industries has been emphasised. As Wu and Ip (1981) observed, planning has, therefore, been successively bottom-up and top-down.

The period 1949–58 witnessed relatively rapid urbanisation. But from 1958, for the next 20 years, the accent was strongly anti-urban and owed much to the ideology of Mao Tse-Tung (Ma, 1976). Urban areas were seen as the hangover of capitalism. After 1958, an extensive rustication programme was embarked upon, whereby youths and urban bureaucrats were sent out to work in the countryside, sometimes permanently. At the same juncture, every effort was made to promote the self-sufficiency of rural communes. The rural areas were encouraged to develop their own resources and to establish their own rural industrial plants. This is a prime example of the basic needs approach associated with increasing territorial closure.

Thus, since 1958, the Chinese have basically renounced urbanisation and migration, and indeed China's 21 per cent level of urbanisation is relatively low when compared with countries of a similar overall level of economic development (Chang, 1982). However, urban growth has continued in China. Shanghai reached a population of 10 million by the mid–1980s, having increased its population from around 5 million in 1949. But today, it is estimated that as high a proportion as one-third of Shanghai's population is engaged in agricultural activities. Certainly, attitudes to urban living have been changed in China and the overall emphasis has been placed on the provision of basic needs and the promotion of social equity. Whether such fundamental policy stances can be explained by ideology alone is highly debatable (Ma, 1976). With well in excess of a million additional mouths to feed each and every month, at one level the transfer of people from the cities to the countryside can be seen as a necessary expedient in order to cope with the ever-increasing demand for food.

Plate 4.1   Restoring part of central Havana after years of neglect (photo: Rob Potter).

Cuba

The socialist state of Cuba provides another interesting example. Prior to the revolution in 1959, Havana stood as the classic primate capital city. The nation was characterised by high rates of rural–urban migration, and the rapid growth of spontaneous settlements in Havana. During this period, it is estimated that around 33 per cent of all homes in the capital were self-built (Stretton, 1978). The Census of 1953 indicated that 55 per cent of all urban housing was either insanitary or inadequate.

Although Castro's reforms were not initially conceived as a communist revolution, he gradually developed them along socialist lines. The leaders of the revolution came to regard Havana as imperialist, privileged and corrupt, and after 1963 the capital was positively discriminated against. No further investment was centred on the capital and the provision of housing and jobs in Havana virtually ceased. This freezing of metropolitan Havana in order to enable the rest of the nation to catch up is frequently referred to in generic terms as the 'Havana strategy' (see Plate 4.1).

At the inter-regional scale, the growth of provincial towns in the 20,000 to 200,000 population range was promoted, in an effort to counterbalance Havana, but in a manner which did not establish new large urban centres. At the next level down, the regrouping of villages into 'rural towns' was effected in order to improve rural living conditions (Hall, 1981b). The government also implemented stringent migration controls, with workers only being allowed to move to new jobs in Havana with ministerial permission. Overall there has been a marked decline in internal migration (Lehmann, 1982).

Although as noted by Stretton (1978) planning in Cuba has frequently tended to be both technocratic and dictatorial, much has been achieved in reducing the urban–rural imbalances which characterised pre-revolutionary Cuba. Promoting social and spatial equity has been an intrinsic part of Cuban development policy. A greater emphasis has been placed on production rather than consumption, and Susman (1987) has argued that the aim of the state has not been to increase consumption per capita, but to increase equity of participation in decisions concerning production. For example, all students are at some stage in their studies integrated into the agricultural process in order to 'overcome attitudes of domination and subjugation', and to reduce elitist attitudes and values in society as a whole.

In the fields of health and education provision there can be no doubting the achievements. Massive efforts have been made to develop new primary, secondary and tertiary health care facilities in rural areas. With respect to education, in 1971 only seven out of 478 secondary schools were to be found in the rural areas. By 1980, 533 out of a greatly enhanced total of 1318 secondary schools were located in the rural zones. It is difficult not to see this as a remarkable attempt to redistribute social surplus product at the national scale.

Clearly, the central aim of policy in Cuba since 1959 has been to 'urbanise the countryside and to ruralise the city'. Reflecting this, the overall level of urbanisation is quite high, currently standing at 60.3 per cent. Although the population of Cuba is still heavily concentrated in Havana, its share of the national total has fallen since 1943. The growth of Havana declined from an average annual rate of 2.7 per cent during 1943–53 to 1.3 per cent in the period 1970–75. Currently, its population stands at the 2 million mark (Plate 4.2).

It remains to be seen what exactly will happen as the result of the emergency period which has been entered since the collapse of the USSR. However, it is hard not to see the Cuban policy on urban and regional development as a major success. Equally, it has to be acknowledged that the price of this reform has been massive and total state control of all areas of human endeavour, that not all have been prepared to countenance, both among those who have endeavoured to flee the country, and among those who write on human rights and development issues.

Grenada

The case of Grenada in the eastern Caribbean is useful in demonstrating that alternative paths to development and change do not have to be revolutionary in the Marxist political sense. In March 1979, Maurice Bishop, a United Kingdom-trained lawyer, overthrew what was regarded as the dictatorial and corrupt regime of Eric Gairy. Maurice Bishop led the New Jewel Movement (NJM), the principal theme of the political movement being anti-Gairyism allied with anti-imperialism. The movement also expressed its strong commitment to genuine independence and self-reliance for the people of Grenada (see Brierley, 1985; Kirton, 1988; Hudson, 1989, 1991; Potter, 1993a, 1993b; Ferguson, 1990).

Plate 4.2   East European-style mass housing on the outskirts of Havana (photo: Rob Potter).

On the eve of the revolution, Grenada suffered from a chronic trade deficit, strong reliance on aid and remittances from nationals based overseas, dependence on food imports and very substantial areas of idle agricultural land. After the overthrow of Gairy, the NJM formed the People's Revolutionary Government (PRG), the movement taking a basic human needs approach as the core of its development philosophy. The PRG stated its intention of preventing the prices of food, clothing and other basic items from rocketing, along with its wish to see Grenada depart from its traditional role as an exporter of cheap produce. The government also set up the National Co-operative Development Agency in 1980, the express aim of which was to engage unemployed groups in villages in the process of 'marrying idle hands with idle lands'.

Between 1981 and 1982, two agro-industry plants were completed, one producing coffee and spices, the other juices and jams. A strong emphasis was placed on encouraging the population to value locally grown produce and local forms of cuisine, although the scale of this task was clearly appreciated by those concerned (see Potter and Welch, 1996). The PRG also pledged itself to the provision of free medicines, dental care and education. Finally, it was an avowed intention of the People's Revolutionary Government to promote what Bishop referred to as the 'New Tourism', a term which is now widely employed in the literature. By 'new tourism' was meant the introduction of what the party regarded as sociologically relevant forms of holiday-making, in particular those which emphasised the culture and history of the nation, and which would be based on local foods, cuisine, handicrafts and furniture-making (see also Pattullo, 1996). Such forms of tourist development, it was argued, should replace extant forms based on overseas interests and the exploitation of the local environment and socio-cultural history.

The salient point is that throughout the period, 80 per cent of the economy of Grenada remained in the hands of the private sector, and a tri-sectoral strategy of development which encompassed private, public and co-operative parts of the economy was the declared aim of the PRG. In this sense, the so-called 'Grenadian revolution' was nothing of the sort. The economy of Grenada grew quite substantially from 1979 to 1983, at rates between 2.1 and 5.5 per cent per annum. During the period, the value of Grenada's imported foodstuffs fell from 33 to 27.5 per cent. Even the World Bank commented favourably on the state of the Grenadian economy during the period from 1979 (Brierley, 1985).

For many it was a matter of great regret that Maurice Bishop was assassinated in October 1983, and the island invaded by United States military forces, because this saw the end to the four-year experiment in alternative development set up in this small Commonwealth nation (Brierley, 1985). This deprived other small and/or dependent Third World states of the fully worked-through lessons of grassroots development that Grenada seemed undoubtedly to be in the process of providing.

Tanzania

The last brief example provided here is that of development planning in the East African state of Tanzania. The area was occupied by Germans in the 1880s, whilst after the First World War it became British-administered Tanganyika. Independence was granted in 1961. Like many former colonies, the population was very concentrated along the narrow coastal region. The other major urban areas such as Morogoro, Iringa and Mbeya formed a corridor running in a southeasterly direction from Dar es Salaam on the Indian Ocean coast. During the era when modernisation thinking was in vogue, 'islands' of development linked by major transport lines were recognised (Gould, 1969, 1970; Hoyle, 1979; Safier, 1969). The settlement pattern traditionally comprised dispersed villages, although strong urban concentration around Dar es Salaam occurred during the colonial period, with Hoyle (1979) referring to it as a 'hypertropic cityport' (see also Hoyle, 1983; O'Connor, 1983).

Lundqvist (1981) identified four main phases of development planning in Tanzania between 1961 and 1980. The period from 1961 and 1966 is seen as the legacy of the colonial era, during which such planning as was carried out was sectoral rather than regional in character, as a result of which, infrastructure remained concentrated in the principal towns and urban–rural disparities were maintained. The main policy efforts to reduce urban–rural differences represent the second and most important phase, extending from 1967 to 1972, which witnessed the emergence of a strong commitment to a rural-based development programme linked to principles of traditional African socialism.

These policies were based on the Arusha Declaration of 1967 which attacked privilege, and sought to place strong emphasis on the principles of equality, co-operation, self-reliance, and nationalism. Such ideas were put into practice in the Second Five-Year Plan, 1969–74. The major policy imperative was *ujamaa* villagisation, which was regarded as the expression of 'modern traditionalism', that is, a twentieth-century version of traditional African village life. The word *ujamaa* is Swahili for familyhood. The intention was to concentrate scattered rural populations, and by this process of villagisation, to provide the services required for viable settlements. Reducing rural–urban migration was a major goal, along with lessening the dependence on major cities such as Dar es Salaam. Ujamaa villages were envisaged as co-operative ventures, by means of which initiative and self-reliance would be fostered, along socialist lines of planning. In addition, efforts were also made to spread urban development away from Dar es Salaam toward nine selected regional growth centres. In overall terms, President Nyerere regarded these policies as a distinct move away from a slavish imitation of western-style planning and development.

But the example of Tanzania also shows the difficulties that can face the implementation of such policies. The third phase identified, from 1973 to 1978, is described as villagisation by order, during which enforced movement to development villages occurred (Lundqvist, 1981; Hirst, 1978). Hirst (1978) attributes this to the speed envisaged for the programme after 1970, which it is suggested almost inevitably meant that coercion would have to be used to compel people to move to new villages. Obviously, this led to the very antithesis of the original aims and intentions of the policy (Hirst, 1978; Briggs, 1983). At the beginning of the phase, in 1973, the centrally located town of Dodoma was selected as the new national capital in place of the peripherally located Dar es Salaam.

Although they have received much praise from certain quarters, the policies adopted in Tanzania have been viewed with considerable scepticism by others, especially those from a committed Marxist perspective. It has been argued that by the fourth stage, starting in 1978, industry and urban development were once again being upgraded at the expense of *ujamaa* villages and rural progress, partly as a reaction to the near-disaster brought about by the phase of enforced villagisation. Thus, the Fourth Five-Year Plan, 1981/82–1985/86, gave priority to industrial development, and by this juncture the *ujamaa* concept appeared to have all but fallen from the consciousness of both planners and politicians alike.

Final comments

This chapter brings to a close our overview of cities and urbanisation in the developing world viewed at the global and national levels. This concluding policy-oriented chapter has highlighted the importance of a number of issues. Firstly, it should be axiomatic that arguments about large cities and urban and regional systems cannot be based on economic reasoning alone. Matters pertaining to social equity and polity are just as important in examining the regulation of urban systems. This is well illustrated in the case of small island systems such as those of the Caribbean, where it may be argued that continued concentration is warranted on purely economic grounds, but where the equity aspects of social polarisation are clear for all to see (Potter and Hunte, 1979; Potter, 1984, 1985; Potter and Wilson, 1990).

The goals of national urban settlement planning are socio-political and the role of the state and its aspirations are, therefore, central to any such activity. Another way of saying this is to stress that the goals of urban systems planning are effectively the goals of society in general. This was well exemplified in the case studies presented at the end of the chapter. National urban development strategies are country-specific and depend greatly on the overall size, and the resource base of the territory concerned. Grenada is clearly in a very different position from China when it comes to implementing the various policy options that are open to developing nations.

Further, even if it is decided that something should be done with respect to the urban system, the seriousness with which policies are pursued is open to a great deal of variation. All too frequently the obvious but vitally important point is missed that decentralisation is a *relative* concept. Thus, limited decentralisation can be promoted within existing national core areas, whilst the status quo is maintained at the inter-regional scale.

Chapters 2 to 4 have also served to emphasise that concentration and inequality have all too frequently been dealt with by means of promoting concentrated development elsewhere. Whilst top-down strategies of development based on the operation of trickle-down effects have been subjected to much criticism in the literature, especially as part of the postmodern critique, many states and global organisations are still following such paths to change and development. Indeed, the World Bank, UNDP and other organisations are returning to the argument that urban–rural growth and large cities are the keys to development and change. They base this argument on the Asian tigers or NICs. This approach will be returned to in several chapters, not least the final account presented in Chapter 10. However, the question that all states should now be addressing, regardless of their political complexion, is on what grounds is it being postulated that growth and development have to be predicated on large cities?

Chapter 5

Urbanisation and basic needs: education, health and food

Introduction

As the previous chapters have exemplified, uneven processes of globalisation have intensified in recent years, with many developing countries experiencing the ramifications of actions taken in other parts of the globe (Johnston, *et al.*, 1996; Daniels and Lever, 1996). The accompanying decline in the power of the nation state, which is a direct outcome of increasing global articulation and neo-liberal policies, raises important questions concerning the ability of Third World governments to control their own development.

Whilst the forces of globalisation are becoming more powerful, the world is also facing a number of key environmental and social problems which require urgent attention. Poverty, AIDS and environmental degradation are just a few of the current issues which face the evolving global order. But whilst international institutions search for revised regulations, over two-fifths of the world's population still lacks access to adequate shelter, food and social services. With national governments becoming less able to pursue autonomous welfare policies under the international call for the privatisation of the public sector, the question of who provides for basic needs is a salient one. The increasing demand for a reduction in state control, combined with the IMF's and World Bank's request for debtor nations to reduce public spending and cut subsidies on food and transport, have left Third World governments less able to implement and fund social welfare projects than they were in previous decades (Stewart, 1995; Allen and Thomas, 1992).

The nature and definition of basic needs

The provision of **basic needs** is seen as a major goal of development policy, but access to adequate supplies of food, health care, education and shelter is rarely given due consideration in the urban context. The **Basic Needs Approach**, which arose following the modernisation paradigm, was initially developed by the International Labour Office (ILO) and was adopted internationally at the 1976 World Employment Conference. Basic needs are directly related to human development and are concerned with providing access to the essential elements of life: food, clothing, shelter, health care (including clean water and sanitation), education, employment and the right to participate in decisions about the future (Stewart, 1985). Since then, the concept has extended to include freedom of speech, self-esteem, human rights and democracy, with gender equality being seen as paramount to the achievement of these goals. Without access to basic needs, 2.2 billion people in the developing world will remain in poverty (UNDP, 1992).

A clear definition of 'basic needs' is problematic, as it is difficult to quantify 'need' in a way that takes into account cultural diversity. As a result, it has been accepted as an ideology and framework for analysis rather than as a precise planning tool. Chambers (1988) notes that there is a 'hierarchy of basic needs', whereby the poor have to satisfy their food needs before enhancing their self-esteem. Feminist analysis has thrown some interesting light on this framework, and has suggested that use of 'entitlements' rather than 'needs' (Sen, 1982) encompasses a wider perspective, and one which recognises that resource distribution is embedded in social and power structures (Kabeer, 1994). Poverty is not caused solely by inadequate 'entitlements' but also by structural inequalities. A suitable framework for analysing poverty, according to Kabeer, is to recognise the interrelation between what she sees as the '**ends**' and the '**means**'. She advocates that attention should be paid both to a household's 'ends', namely providing basic needs, and its control over the 'means' by which these are to be met (Kabeer, 1994: 143). It is often the 'means' which are subject to social and political control, and which are strongly influenced by gender, ethnicity and class.

More recently, and partly due to the fact that structural adjustment policies mean that poverty is expanding to incorporate a new group of the urban poor, drawn from the lower middle classes, there has been a move to reconceptualise poverty and basic needs (Amis, 1995; Stewart, 1995). It is difficult to define urban poverty precisely, and attempts to measure it have usually overemphasised economic and financial considerations at the expense of social and cultural ones. As a result, Wratten (1995) argues that urban poverty is likely to have been underestimated in recent decades.

Attempts to redefine poverty have included the UNDP's '**capability poverty measure**' which assesses households' ability to cope at times of crisis (UNDP, 1996). Following Sen's research on entitlements, attention has also turned to analysing the nature of a household's assets, which include its human capital (skills and education), its household facilities, and its **social capital**. Social capital, a concept derived from the work of Putnam (1993), refers to the relationships and levels of trust that people have both within their own community and outside it. Moser (1996: vi) notes that social capital 'is the trust, reciprocal arrangements, and social networks linking people in the community'. In a recession, a person with more social capital should be able to call on more favours and assistance, which should enable them to cope more effectively with poverty (Evans, 1996).

Social capital and networks are extremely important in the informal labour market, as labour power is seen to be the poor's greatest asset. As a result, attention needs to be focused on a household's **vulnerability**, which represents the level of insecurity and lack of well-being of a household and its ability to cope with change, rather than its level of poverty *per se*. The concept of vulnerability is rapidly being accepted as a useful tool for analysing the position of the urban poor, because it takes into consideration a household's non-financial assets (Satterthwaite, 1995; Moser, 1995a, 1996).

In its simplest form, then, the provision of basic needs is recognised as the best means to alleviate poverty at the household level. The provision of basic needs in urban contexts is made problematic by the rapid growth of cities and their populations, which is often beyond the carrying capacity of existing systems and resources. As economic and social change occurs, with increases in contract employment and agricultural modernisation in rural areas, people will continue to move to cities. At the very time when demand for housing, sanitation, education and health care is ever increasing, governments' ability to provide resources has generally been diminished. On the one hand, global institutions are promoting the concept of basic needs for all, whilst on the other, agencies such as the IMF and the World Bank are demanding a reduction in public spending on social programmes as part of the process of structural adjustment (George, 1988; George and Sabelli, 1994).

It is well known that the propensity to meet basic needs is greater in the city than in smaller urban centres and rural areas (Drakakis-Smith, 1996b). As reviewed in Chapters 2 and 3, urban bias (Lipton, 1977; Chambers, 1983) has resulted in the unequal distribution of welfare services, shelter and employment between cities, often with an over-concentration in the nation's capital, where foreign investment, wealth and the elite classes are concentrated. Contrary to much misfounded speculation, at its best the quality of education and health care available in Third World cities is frequently similar to that found in industrialised countries, but access is often restricted to the wealthy, as noted in Chapter 1. The present chapter will stress that it is equal *access* to basic needs rather than *provision* which is the most pressing issue in the urban context.

Within developing cities, as elsewhere, access to basic needs is governed not just by rational economic choice and availability of resources, but by social and political processes which result in unequal access among different groups of the population. The view that city life diminishes differences based on ethnicity and class is misfounded, as cities all around the world are facing tensions caused by the intensification of differences between residents in times of recession. In reality, individual access to basic needs is not just based on the ability to pay, but often determined by the interactive social and economic forces which may be embedded in historical, political and religious factors.

If, as Simon (1981) has suggested, people are the ultimate resource, then a city's development is closely related to the welfare of its population. An educated and healthy workforce is better able to participate in the fight for democracy, social justice and equality, as well as contribute to economic development. In order to illustrate the complex relationship existing between welfare and urban development, this chapter discusses the provision of, and access to, basic needs in the developing city. By drawing on specific examples, it focuses on education, health and access to food, whilst housing and employment are discussed in Chapters 7 and 8 respectively. A primary focus in the present chapter is the level of social inequity in the city as a result of current political, economic, social and cultural structures, all of which are affected by processes of **global change**.

Accordingly, the chapter is divided into four sections. Firstly, in order to appreciate the nature of social stratification in the city, a brief synopsis of gender, ethnic and class constraints is provided. These dimensions are not always given due attention in the literature. As noted previously, the chapter shows that it is not always the provision of basic needs that is inadequate, but rather that individual and community access to such needs is highly unequal. In this respect, it is shown that a **gendered approach** to urban development is crucial, not least because women often suffer disproportionately from urban poverty (Chant, 1996a). This is not only as a result of social divisions and cultural practices, but due to an in-built male bias in the development process (Elson, 1995), whereby the outcome of gender-blind development policies often adversely affects women (Moser, 1993; Young, 1993).

Whilst it is not possible to provide an in-depth review of all aspects of the literature on gender and development, this chapter will attempt to draw on some of the more salient points, and to highlight the importance of adopting a gendered approach to the analysis of urban development. Following on from this, the account moves to discuss education provision before analysing urban access to health, nutrition and food. The final concluding section ends on a positive note by analysing recent social change involving the mobilisation of the poor across many cities. The conclusion tentatively suggests how the processes of globalisation may have provided positive assistance to some communities in their quest for equality and empowerment.

Social stratification and access to basic needs: gender, ethnicity and class

As noted in Chapter 1, urbanisation is a social as well as a structural process, which brings with it new forms of social organisation and new cultural experiences. City living often differs markedly from rural existence, and not just in terms of lifestyle and opportunity. City residence frequently involves a closer relationship with political institutions and national governance. Laws and regulations, as well as unwritten codes of acceptable behaviour, are more likely to be enforced in cities, which frequently exemplify more extreme forms of social control. The nature of this control differs from city to city, but is likely to be heavily influenced by politics, religion and colonial history. For example, in countries where Islam is promoted by the state, religious beliefs govern the legal and institutional framework (Afshar, 1991, 1996). Under the apartheid regime, race and colour affected every aspect of urban life, from access to resources to physical movement around the city (Smith, 1986). For much of the developing world, cities are socially and culturally embedded in a colonial history whereby European traditions and codes of behaviour were superimposed on existing social structures. As a result, segregation in the city is often based on gender, ethnicity and class.

Urbanisation, poverty and gender equality

Poverty and urbanisation are highly gendered, with women constituting the poorest of the poor in most developing cities. The increasing poverty of women witnessed in the last 20 years has been attributed to their unequal position within the labour market, their subordination under patriarchal social systems, and the undermining of their status and power through capitalism, all of these factors being exacerbated by structural adjustment and global economic trends. According to Seager and Olson (1986) women undertake two-thirds of the world's work, provide 44 per cent of the world's food, but earn only 10 per cent of world income and own only 1 per cent of all property.

For years, the *Human Development Report* has identified the basic aim of development as 'enlarging people's choices' (UNDP, 1995: 1). Paramount to this objective is equal access to human rights and basic needs for both women and men, a goal which has still not been achieved universally. The Vienna Declaration, adopted in 1993, outlined the many dimensions of human rights for women, and the need to improve their access to social services, political decision-making, equal work and reward, and protection under the law. The 1995 *Human Development Report* was devoted to 'the relentless struggle for gender equality' based on the argument that 'human development, if not engendered, is endangered' (UNDP, 1995: 11).

Gender equality, however, is not a goal but a process which challenges established notions of social and economic structure, including stereotypical images of women's and men's position in 'development'. As the growing body of literature on gender and development argues, development is often subject to male bias (Elson, 1995; Moser, 1993; Kabeer, 1994; Young, 1993). Conventional development theories are all too frequently inadequate in terms of providing a basic framework to explore gender. Most 'theories' of development have little to say about women, for in the past, research and policies have generally been directed towards men. Women's exclusion from development was not challenged until the 1970s, when Boserup (1970) argued for the recognition of women's productive roles in Africa. Following this, the 1975 International Decade for Women brought attention to women's subordinate position in many societies.

Broadly defined, the gender and development policies which have ensued can be identified as belonging to one of three approaches: Women in Development (WID), Women and Development (WAD) and Gender and Development (GAD). **Women in Development** policies, which were firmly rooted in the 1970s, are concerned with welfare and anti-poverty approaches. Until recently, it was believed that economic growth and development would automatically 'trickle down' and improve women's status as a matter of course. Moser (1989a) argued that these policies were fundamentally flawed, as women were seen solely as reproducers, a factor which owes some of its foundation to notions of the 'western housewife' (Buvenic, 1983; Rathgeber, 1990). Where women were included, they were often seen as *objects* rather than as *agents* of change.

Women and Development approaches are more concerned with improving gender equity and the 'efficiency' of development policies, and are based on the premise that development will not be 'efficient' unless women are included. These approaches are particularly popular with international agencies such as the World Bank, unlike the third and most recent approach which is favoured by western and Third World feminists alike. **Gender and Development** is the latest and most radical approach which challenges current global social and economic structures. It argues that development policies should not only assist women in acquiring **Practical Gender Needs (PGNs)**, but should enable them to empower themselves and confront oppression through **Strategic Gender Needs (SGNs)**, involving grassroots mobilisation (Moser, 1993; Kabeer, 1994).

Although there is a wealth of literature and excellent research on the gender implications of development, global institutions and agencies have largely failed to take a gendered approach in any meaningful way. As Moser (1993) argues, a true gendered development challenges the entire economic and political system and questions capitalism itself, hence its lack of popularity outside women's organisations and academia. It has, however, been widely pursued by Oxfam through their Gender and Development Unit (GADU), and other women's NGOs. A good example of a GAD approach would be assisting women to organise their own health care clinic, which would also afford a forum for discussion and education (Macdonald, 1994).

Urbanisation often intensifies societal notions of men's and women's role in the city and may promote gender differences for a number of reasons. Firstly, in contrast to rural livelihoods, urban households are perceived to be more reliant on waged labour than subsistence production, whereby men take greater responsibility for production. As Chapter 8 discusses, however, a great deal of valuable subsistence and reproductive work is undertaken by women in urban households, but it is more 'invisible' and undervalued than in rural communities. As a result, women's economic and reproductive contributions to urban society are frequently not accounted for, a situation which undermines their power both within and outside the household (Massiah, 1993; Stitcher and Parpart, 1990; Scott, 1994).

Secondly, reproductive activities are often seen as women's work. Women's responsibility for reproduction and household duties has been identified as the prime reason for their disadvantaged status in most societies (Beneria and Sen, 1981). Thirdly, urban women may find that their gender roles are tightly linked to social attitudes and codes of behaviour related to class or religion. Images of appropriate social behaviour for women are often accentuated in urban life, as notions of public acceptability are more widely adhered to. As a result, men and women may use the city in different ways, and for different purposes. Thus, the city is frequently spatially divided by a number of **gendered** and **racialised boundaries**, which subtly control individual access (Lloyd-Evans, 1994). Particular ideas relating to women's and men's use of space serve to influence where people work, live and conduct their daily lives (Hanson and Pratt, 1995). Lloyd-Evans (1994) found that working in certain parts of Port of Spain, Trinidad, was not seen to be socially acceptable for female informal traders. As a result, female traders were generally prevented from working in the more profitable streets of the city. Notions of women's use of 'public' as opposed to 'private' spaces, such as the household, have been well documented (Redclift and Sinclair, 1991). In this way, urbanisation is inextricably linked to social and cultural processes which may undermine the ability of households to meet their basic needs.

Gender roles are also affected by external, as well as internal, economic and social processes. External forces, such as structural adjustment and globalisation, have affected women's ability to meet their basic needs. The gender division of labour in many cities is such that women are frequently engaged in low-paid, exploitative and arduous work, whilst also maintaining the household (Chant, 1996b). In addition, the transfer of state responsibilities to the community increasingly places a further burden on women. The 'triple burden' of reproduction, production and community participation stretches women's ability to cope with a crisis (Moser, 1993). There is, however, an inherent danger of seeing all Third World women as an homogeneous group of passive, poverty-stricken, oppressed workers that need help (Mohanty, 1991). In this respect, the concepts of race and ethnicity, as well as class, are also extremely important in understanding oppression and poverty within the city.

Social difference in the city: class, race and ethnicity

The concept of 'difference', particularly in relation to gender and ethnicity, has been addressed in the geographical literature (see Carby, 1983; Jackson, 1987; Westwood and Radcliffe, 1993; Maynard, 1994), but not widely in empirical research undertaken on the developing city. As McIlwaine (1995: 239) argues, 'whilst Third World women's voices contributed to theoretical orientations within the social sciences, these have not yet been followed through in terms of addressing both gender and race within the formulation of development policies'.

For many women and men, problems of nationality, class and race are all linked to oppression. Cleves Mosse (1993) talks of gender, race and class constituting a 'triple yoke of oppression' for many women, particularly in post-colonial societies. It is necessary to understand more about the legacies of hierarchy and inequality left by colonialism, and to question the role of many post-colonial institutions which serve to structure relations in the modern world. Race and ethnicity are important components affecting social differentiation in most developing cities, as they involve subtle forces of discrimination. Squatter settlements, for example, are far from socially uniform, being comprised of individuals originally drawn from different regions and distinct ethnic backgrounds. Individual settlements are often internally divided along distinct class, ethnic, linguistic and religious lines.

Although ethnic divisions are seldom as extreme as those of South Africa's apartheid regime, and do not lead to the extreme violence witnessed in Sri Lanka, they are nevertheless extremely important. In most nations, particularly

those of Latin America, Africa and the Caribbean, occupational status and social hierarchy are closely interlinked with colour. The elites are frequently descendants of the old European sector (Gilbert, 1990), and there is little social
mobility between classes. In Indian cities, however, there are reports of greater
social fluidity for lower castes due to improved employment opportunities.

Cubitt (1995) refers to the Latin American 'colour–class system', which is
based on the historical interaction between indigenous American, European
and Black African peoples. Radcliffe and Westwood (1993) note that indigenous ancestry has been negatively perceived in Latin America, with much
discrimination against Indian descendants in society. Radcliffe (1996) reports
that impoverished indigenous female migrants provide labour for urban centres, and argues that this domestic service example 'highlights how the spatial
and social relations based on race, gender and generation simultaneously and
inextricably locate individuals in Latin American societies' (Radcliffe, 1996:
162).

In the Caribbean, there exists a hierarchy of colour, with 'whiteness' associated with higher social status. In Trinidad, colour, wealth, education and
culture dynamically interact as markers of differential social status. The literary
scholar C.L.R. James (1963: 83) noted that the living standards of Trinidadians
are equated with their group's standing in a 'pluri-defined social totem pole'
(see also Yelvington, 1993; Lloyd-Evans, 1994). 'Playing for white' is a common term used amongst the aspiring black middle classes. Ethnic stereotyping
in relation to education, employment and social behaviour perpetuates differences and maintains segregation in the labour force. In this way, vulnerability
is determined by a number of social, economic and political forces. Increasing
access to basic needs has been identified as one way of enhancing the social
mobility of marginal groups. Education in particular has been seen by many as
a path towards poverty alleviation, but as will be argued in the next section,
increasing provision of schooling will not necessarily serve to challenge prevailing social structures.

Education in the city: a path to social mobility or social inequality?

> Among the developing world's 900 million illiterate people, women outnumber
> men two to one. And girls constitute 60 per cent of the 130 million children
> without access to primary school. (UNDP, 1995: 4)

Education has an important role to play in development, in terms of both
structural change and individual opportunity. The need for education is undis-
puted in the debate on development, although there is contention over its
quality and content. Education was recognised as a basic need by the United
Nations in the 1950s, and it is a key component of equitable development.
Education is usually associated with the provision of formal schooling and
has essentially been regarded as the responsibility of the state, although in the
1990s this sector was not excluded from the process of privatisation.

Throughout the developing world, education is seen as the key to individual advancement, as well as a key contributor to national social and economic development. Poor nations have invested vast public revenues in education, as governments and individuals alike believe that education is a cure for the 'ills of development'. On the supply side, governments and international agencies such as UNESCO and the World Bank have recognised the need for an educated population, whilst schooling is increasingly demanded by the poor themselves, who see it as the only stepping stone to social and economic advancement.

From a national viewpoint, a literate workforce will be better equipped to deal with innovations and technical change. Secondary school and university graduates are needed to assist in the country's professional development and reduce dependency on outside expertise. However, as Gould (1993) argues, the notion that education alone will open the doors to a better standard of living and to enhanced social mobility is perhaps one of the unchallenged myths of development, as like any other resource, its quality and distribution are governed by economic, social and political forces. Whilst not disputing the benefits of education for poor communities, it is important to recognise the fact that educational systems can be developed to maintain, rather than challenge, the unequal social hierarchies previously discussed. This argument parallels some of those made in relation to regional development and the generation of inequalities in earlier chapters.

In 1990, the UNDP reported that 36 per cent of adults in the developing world were illiterate, with sub-Saharan Africa accounting for the highest rates of illiteracy (50 per cent), and Latin America the lowest (16 per cent). However, education provision has improved greatly over the last 30 years (UNDP, 1995). Between 1960 and 1990, the total number enrolled in the three main levels of education in the developing world rose from 163 to 440 million (Todaro, 1995). The most marked increase has been seen in primary school enrolment, which accounts for 78 per cent of total enrolments in the developing world.

Despite the undoubted gains made in global education, statistics do not reveal the proportion of children who finish school, nor do they measure the quality of education attained, and there is much regional variation. In Latin America, for example, 60 per cent of children who enter primary school drop out before completion. Even for those children who do complete their primary education, secondary schooling is generally very expensive and beyond the reach of most poor families. By contrast, university education is frequently heavily subsidised by the state, but without free secondary education the poor are effectively prohibited from access to professional careers (Gould, 1993; Graham-Brown, 1991).

Local education systems are invariably linked to the global system, in respect of both policies and rhetoric. From the fifteenth century onwards, education was seen as a tool of civilisation in the developing world, and it spread with colonialism. The first school for Indians in Latin America was founded by Franciscans at Texcoco (Mexico City) to enable the indigenous population to learn trades. In the fourteenth century, when class, race and gender determined access to learning, Spain contributed high quality universities to Latin American cities such as Mexico City and Lima (Scarpaci and Irarrazaval, 1996). Today, most ex-colonies have an education system based on that of the colonising power, with the church retaining its powerful influence. Church-run schools are extremely important in many African and Latin American cities. In Latin America, schools and colleges are largely part of the church's 'empire' which allows the diffusion of particular values and codes of behaviour pertaining to the Roman Catholic religion (Cubitt, 1995). Christian missions still play a major role in countries such as Zaire, which raises questions relating to the promotion of neo-colonial views through education. Dixon (1992) argues that education is still an important factor in the process of 'internal colonialism' which leads to the suppression of indigenous cultures.

In Africa, indigenous forms of education involving community development are still extremely important, but are undermined in a global system where education is synonymous with formal schooling (Bray et al., 1988). In many Islamic areas, children attend schools to learn the Koran, and this may be their only experience of school. In southeast Asia, Dixon (1992) states that national education schemes are used to impose uniform nationalist views on the population. However, education systems which are based on European curricula and which promote neo-colonial tendencies are increasingly being questioned (Gould, 1993). The teaching of children in a colonial language, for example, is contentious in many Andean and African cities where native languages are prominent. Further questions relate to whether the distribution network of knowledge and learning has to be via the formal education system.

'Human resource development' is a populist term of the 1990s, and with it comes a new ethos surrounding the nature of education. Whilst it is widely accepted that education and training are important foundations for social and economic development, the provision of education in cities is markedly unequal. So, given the growth of the education sector across the developing world, what benefits have urban societies accrued? Has better education aided social mobility in the city, or as some authors suggest, has it merely reinforced the existing social hierarchy? Before addressing these important questions, it is necessary to analyse the spatial distribution of educational services and the processes which control individual access to those resources.

Educational provision in the city: one step forward, two steps back?

The global support for education, which cuts across different political and social regimes, has resulted in a huge increase in the provision of primary school places in developing countries since the 1950s. This expansion was partially fuelled by the many newly independent states in Africa, Asia and the Caribbean, which fostered education as a way of promoting social and economic equality. The widespread increases in public expenditure on education in the 1960s and 1970s exceeded increases in any other sector of the economy (Todaro, 1995).

The systematic increase in the level of education in Africa, Latin America and Asia has been remarkable. In tropical Africa, 70 per cent of children now attend primary school with the more urbanised countries of Nigeria, Zimbabwe and Kenya attaining the highest levels, although countries like Ethiopia and Somalia still only record around 25 per cent (O'Connor, 1991). As will be discussed, however, national enrolment rates in Africa are much lower for girls than for boys, particularly at the secondary level. Progress in southeast Asia has also been considerable, with only Laos and Cambodia reporting less than 100 per cent of the eligible 5 to 11 age group in primary education (Dixon, 1992: 135). Despite similar efforts to promote secondary education, only in the Philippines, Singapore and Malaysia do more than 50 per cent of children aged 12 to 17 attend school.

The global expansion of schooling has meant that most children in the developing world now attend school at some time in their lives. In Latin America, CEPAL (1991) reported that the regional average for secondary school enrolment rose from 19 to 49 per cent between 1950 and 1980, with the greatest increases having been experienced in Chile, Peru, Ecuador and Colombia, although there has been an increase in drop-outs following the external debt crisis. By the end of the 1980s, stagnant economic growth and structural adjustment forced many governments to reduce their spending on social programmes, including education. Many would argue that increasing social and economic problems in the 1990s have all but undone the educational progress made by Third World countries in the previous decades. Despite widespread policies, generations of children are still growing up without the necessary literacy and skills needed to gain good jobs. Furthermore, many schools have poor facilities such as no latrines or water, and are not conducive to learning (Satteur, 1993). Instead, too many urban children obtain their 'schooling on the streets', engaging in work or begging, topics which receive further attention in Chapter 8.

Education is inextricably linked to city growth and development. Firstly, there are strong links between education and migration at a range of spatial scales (Chant, 1992; Gilbert, 1990). For the professionally qualified, overseas opportunities in the west are often more attractive than remaining at home. The 'brain drain' has reduced the availability of doctors, lawyers and other professionals to assist with city development. Western-oriented learning at European and American universities for the elite classes has all too often failed to equip return migrants with the skills needed for appropriate local development initiatives. The moderately educated, however, are more likely to migrate to cities from rural regions for better employment and welfare opportunities in, for example, the burgeoning public sector.

The search for higher education in particular leads to inter-urban as well as rural–urban migration. According to Gould (1993), education is a prime cause of migration to Karachi in Pakistan. The 'internal brain drain' which results in the movement of the educated to the city, whilst the illiterate stay in rural areas (Lipton, 1980), is a reality in many countries. Without rural development and employment provision, this flow of educated, skilled labour will continue to contribute to inter-regional disparities. Women, in particular, constitute a large proportion of rural migrants, as the formal labour market for women is concentrated in urban areas (Chant and McIlwaine, 1995b). In Nairobi, 40 per cent of the population between 15 and 40 have been to secondary school (Gould, 1986), and in Indonesia, the internal loss of population to Jakarta has resulted in an over-supply and the de-skilling of many jobs. The 'diploma disease' (Dore, 1976), or the over-supply of educated workers, has led to the downgrading of jobs which require degrees.

Secondly, educational provision is greater in cities than in more peripheral or rural areas. In particular, higher education is almost exclusively located in primary or secondary cities. Albornoz (1993) states that urban schools in Latin America are better resourced than those in rural areas, a fact which holds for many nations. Despite great gains, public demand for education still exceeds supply and this situation is unlikely to change. Scarpaci and Irarrazaval (1996) argue that in Latin America and the Caribbean, public education is generally of low quality, although provision is high.

Education for whom? Social inequality and access in the city

Despite marked improvements in the provision of school places, access to schooling for the urban poor is still markedly unequal as a result of two factors, income and gender. In many lower-income communities across the globe, primary education is provided at some cost to the household. Even when schooling is free, teaching resources such as books, equipment and compulsory uniforms are rarely provided. Furthermore, many peripheral squatter settlements have no access to a nearby school. In a study of a poor urban community in Guayaquil, Ecuador's largest city, Moser (1989b) found that although state schools were 'free', a number of fees had been imposed on parents during the 1980s crisis. By 1988, the costs of a child attending school, including uniform, books and transport, amounted to between one and two minimum salaries. Thus, many poor families work to send one member through school, usually a boy, in the hope that they will support the family in later life.

Although cities have the highest levels of expenditure on education and relative provision of school places, the huge size of many Third World cities and their youthful populations mean that access to even primary education is severely restricted. In Peru in 1990, the newly elected President Fujimori introduced a severe adjustment programme designed to cut spending on education from its 1980 level of $62.50 per head to $19.50. Teachers' wages were reduced to a quarter of their previous levels, which resulted in mass exoduses of teachers and students (Green, 1995). Indeed, in all too many societies, teachers are poorly paid and exist on the poverty line themselves.

Under conditions of austerity, many cities rely on the private provision of education to meet the growing demand. In many countries, good secondary education is mostly available to the middle classes who can afford it. In 1990, one out of every two poor children were behind in their education in Brazil's cities, compared to one out of 10 of the richest 25 per cent of the urban population (Green, 1995). In Mexico City, for example, wealthy parents can choose to send their children to a number of good private secondary schools depending on whether they favour the British, American, French or Mexican systems of education. Access by poor students to university is limited due to the high cost of secondary education and the opportunity cost of their labour. Although universities generally command lower fees, entry to the National Autonomous University of Mexico (UNAM) in Mexico City is often through social selection rather than academic ability, something that is also true of the city's key professional jobs. According to Cubitt (1995), UNAM brings 400,000 students into Mexico City, but drop-out rates are high due to high costs and lengthy degrees.

Gilbert (1990) argues, however, that the expansion of secondary and university education has benefited the middle classes, and this in turn has altered the social structure of many Latin American cities. He argues that the 'middle sector' now includes one-fifth of the urban population in the more prosperous cities. In contrast, Filguera (1983) states that whilst education in Latin America is increasing, the gap between rich and poor is widening. In this instance, increased provision of education has done little to enhance the social mobility of the poor or to redistribute wealth and power, and in particular, there has been no corresponding reorganisation of the urban labour market to provide more equal access to jobs for the educated.

Access to education is also highly gendered (Ballara, 1992), and often politicised or linked to religious practices which are biased against women. Although the proportion of girls in primary education has risen globally, the poorest nations in particular have lower female enrolment levels. The difference in enrolment rates for girls is most marked in the Middle East and North Africa, whilst female literacy is 31 per cent lower than that for males in the developing world taken as a whole. According to the UNDP (1995), however, the rate of female enrolments in primary school as a percentage of male enrolment has risen from 61 per cent in 1960 to 91 per cent in 1988. However, secondary level education is more highly differentiated. UNICEF estimate that twice as many boys as girls are enrolled in secondary school in half the countries of tropical Africa (UNICEF, 1995). According to Young (1993), education for girls is not enough and not the right sort.

The NGO 'Womankind' identifies three barriers to education for women: the global economy, social attitudes and the school environment itself. Until the 1980s, school enrolments for girls had risen substantially, but reduced dramatically under structural adjustment, particularly amongst female-headed households. Restrictions in household incomes are likely to result in girls being taken away from school before boys, a factor which is related to ingrained social attitudes about the role of women. As a result, women lack the resources which would allow them to cope more effectively with poverty and the effects of structural adjustment. In Mexico City in the 1980s, a child could not enrol at school without a birth certificate which named the father. This forced recognition of parentage has led to many children being denied their right to education. In Mexico, illegitimacy is stigmatised even though it is common amongst poor communities. For many single mothers this presents a problem, not least because of the expense of registering a child's birth.

Cleves Mosse (1993: 81) argues that education for girls is often a double-edged sword, for it can serve to reinforce notions of female inadequacy through the curriculum. In many cases, increasing educational opportunities for women has done little to improve their welfare and job chances later in life. There is also a lack of understanding over the conflicting demands made on girls between school and home.

Potential solutions which will assist in opening the paths to education for all city dwellers are varied. On the one hand, formal education is seen as the way forward, but given its high cost, it is unlikely to expand under current economic conditions. The *1990 World Conference on Education for All* declared that an expanded vision was needed to promote education globally. The public funding of education in cities is likely to decrease as liberalisation is encouraging a transfer of state provision of education to the family and community, which has far-reaching consequences for marginal groups. Although there is already much community involvement in school provision, often funded by NGOs, their success is dependent on a constant supply of resources. Furthermore, many urban communities are built on their own gender and class hierarchies, which maintain inequality at the local level.

Another debate centres on whether education should be synonymous with formal schooling. Although literacy is essential, it could be argued that more vocational or indigenous forms of education might equip poor children better for the future. Whilst this approach has its merits, particularly in rural areas, it may not bring as many advantages in the city where access to well-paid stable employment is through the formal education system. Urban children with alternative skills and non-formal education will ultimately lack access to better jobs in the formal sector. The dilemma is such that training children to undertake pottery or metalwork in a poor urban community enables them to make a living in the informal sector. However, it can also be said that it reproduces the social class system, as those children are prevented from social mobility through employment.

Within cities, increasing technological innovation and industrialisation will require a more skilled labour force. In this sense, education can be viewed as a tool of global capitalism, seeking to produce adequately skilled workers for foreign multinationals. In the past, socialist governments have attempted to restructure education to indigenous needs and principles, as in President Nyerere's promotion of 'education for self-reliance' in Tanzania in the late 1960s. It is pressing that national governments should adapt their education systems to the needs of their populations. In the urban context, this does not necessarily mean abandonment of formal schooling, but a move towards a more indigenous and culturally appropriate system. In Trinidad, a national youth training scheme, the Youth Training and Employment Partnership Programme (YTEPP), assists school leavers in developing vocational skills which are suited to small business and informal employment (Lloyd-Evans, 1994, 1995).

Even with improved educational provision, many children are unable to benefit fully from the opportunities available as they suffer from ill health, hunger and malnutrition. Education policies alone will be unsuccessful if other basic needs are not met at the same time. Poor health and hunger, in particular, are often 'invisible' urban problems which undermine children's future life chances. The challenge thus remains a very substantial one indeed.

Urban health, food and malnutrition: the hidden epidemic _____

In the early 1980s, the world's media brought horrific images of hunger and famine in Africa to households around the globe. This vivid portrayal of death and human suffering promoted much concern over the causes and consequences of hunger. Whilst television documentaries revealed the fact that famines were as much the result of political issues and human-induced factors as the outcome of drought and climate change, western perceptions of hunger identified the problem as a predominantly rural one.

As the world sympathised with Ethiopia's children, another hunger-related epidemic was contributing to the death of children each year in cities right across the developing world, namely malnutrition. Due to the warnings about 'urban bias' in development and the over-concentration of the wealthy classes in Third World cities, statistics for infant mortalities, disease and malnutrition usually identify rural populations as those most at risk. Although medical service provision is primarily concentrated in cities, as in Guatemala where 80 per cent of medical services are located in the capital (Painter, 1987), the majority of the urban poor still have highly restricted access. As Foster (1992: 1) argues, 'Most hunger related deaths do not occur in famines . . . but daily, quietly, largely unchronicled, all around the world'. This chronic level of urban hunger and illness is often invisible, and is hidden amongst positive statistical indicators.

The FAO estimate that 10 million die annually from hunger, whilst 12 per cent of the world's population is hungry. It is also a condition that impacts on different sectors of the population unequally, discriminating according to age, lifecycle stage and gender. In the last decade, due to economic recession and structural adjustment, infant and maternal (pregnant mother) mortality rates in many Third World cities have increased, whilst many previously combated diseases like tuberculosis are spreading at epidemic rates. Furthermore, as developing cities diffuse western culture through unregulated advertising, so smoking and bad eating along with drug abuse have increased the incidence of heart disease, lung cancer and respiratory failure. Improving urban health care is a complex and multi-faceted task (WHO, 1992; World Bank, 1993b; Stephens and Harpham, 1992). It is not just about the availability of medical staff or basic drugs alone, but involves access to a combination of basic needs including shelter, clean water and sanitation, together with a stable income. Cubitt (1995) argues that the poor lack access to health care as a result of economic, cultural and institutional factors.

Table 5.1   General trends in development in a selection of developing countries

Country	Infant mortality (‰)		Population with safe access to water (%)		Underweight children under the age of five (%)	
	1960	1992	1960	1992	1960	1992
Hong Kong	44	7	99	100	na	na
Costa Rica	85	14	72	93	10	8
Chile	114	16	70	86	2	2
Trinidad and Tobago	56	18	93	97	14	9
Mexico	92	36	62	84	19	14
Thailand	103	37	25	77	36	13
Brazil	116	58	62	87	18	7
Peru	142	64	na	na	17	13
Philippines	80	44	na	na	39	34
South Africa	89	53	na	na	na	na
Zimbabwe	110	67	na	na	25	14
Pakistan	163	91	25	68	47	42
Kenya	124	69	17	49	25	17
India	165	82	na	na	71	63
Mozambique	190	148	na	na	44	47

Source: UNDP, 1995

Health and the urban environment

In a pattern which is similar to education, the systematic progress made in health care provision over the last 40 years, particularly in respect of immunisation, the increase in life expectancy and reduction in infant mortality, has been well documented (UNDP, 1992, 1995; Phillips, 1990; Phillips and Verhasselt, 1994). As indicated in Table 5.1, rapid improvements have been made in certain Asian, Latin American and Caribbean countries through the implementation of low-cost national projects which concentrate on basic drug and health care provision. Despite evidence of progress, South Asia and Africa still have high rates of infant mortality. Health care provision has in fact followed a similar pattern to the expansion of education in recent decades. Major improvements were recorded during the period from 1950 to 1980, when infant mortality rates decreased rapidly as a result of government expenditure on immunisation and basic services. In the 1980s, this progress was halted and in many cases undermined by global events. The 1980s also witnessed the start of the global AIDS epidemic which has had far-reaching consequences for Third World nations.

The 1980s saw a polarisation of health and nutrition between social groups, particularly in Latin America, where the debt crisis was most severe for urban inhabitants. After nearly three decades of substantial improvements, health and nutritional conditions deteriorated sharply in the majority of Latin American and African countries, whilst only in Southeast Asia did they continue to improve. Real expenditure on health declined between 1979 and 1984 in 47 per cent of African, 61 per cent of Latin American, 43 per cent of Middle Eastern and 33 per cent of Asian countries (Cornia, 1990). At the same time, expenditure on basic food subsidies declined whilst the real wages of the poor fell.

The provision of health care by the state varies across the developing world. In Latin America, only Cuba and Nicaragua maintain national health systems that are governed by the Ministry of Health. In a large proportion of countries, including Brazil, Chile, Costa Rica, Mexico and Venezuela, health care is dependent on insurance systems, in either the private or public sectors. Qualifications for insurance and social security include a full-time job, which effectively excludes a vast number of informal workers. Scarpaci and Irarrazaval (1996) further differentiate between Latin American and Caribbean countries according to whether their health policy is 'permissive'/*laissez-faire* (Argentina, Uruguay, Venezuela, Bolivia, Chile, Paraguay and Honduras), 'co-operative' between public and private resources (Brazil, Jamaica, Mexico, Trinidad and Tobago, Colombia, Costa Rica, Ecuador, Guatemala, Peru, El Salvador, Haiti, Nicaragua), or 'socialist' (Cuba).

In Africa, where a large proportion of the population suffers from illness for their entire lives, there are both indigenous as well as western-style systems of care (O'Connor, 1991). Again, money is often spent on increasing the number of highly trained doctors and high technology equipment in a few capital cities, rather than on primary health care. AIDS, although not exclusively an African disease, has added to the hazards faced by many urban populations. The World Health Organisation (WHO) estimates that by 1990, 3 million people in Africa were infected with the HIV virus. The virus, which is particularly intense among the young, has the potential to damage national economies as well as individual lives. About 10 per cent of cases are found in very young children who have been affected by their mothers (Barnett and Blaikie, 1992). For cities, such problems have been exacerbated by the rapid process of urbanisation and environmental degradation.

Pryer and Crook (1988) have argued that the combination of deteriorating urban environments, inadequate shelter, increased poverty and reduction in government spending has seen health conditions deteriorate faster in Third World cities than in the surrounding rural areas. Cholera, which had been wiped out in the developed world, reappeared in Peru in the 1990s and spread through Latin America to Mexico (Cubitt, 1995), largely as an outcome of poverty. Increased industrial pollution and environmental degradation, overcrowding in squatter settlements and slum tenements, inadequate water and sanitation, have added to malnutrition and the risk of infectious diseases. The main health problems experienced by the urban poor include birth-related ones, digestive infections, infectious disease (both vaccine preventable and non-vaccine preventable), respiratory infections, insect-borne diseases such as malaria and yellow fever, and intestinal parasites (Hardoy *et al.*, 1992).

As Table 5.1 indicates, infant mortality is still a major problem in many Third World countries. Pryer and Crook (1988) estimate that up to one-third of children in developing nations die before they reach the age of five. UNICEF (1995) report that tetanus, diarrhoea, measles, malaria and respiratory infections account for 90 per cent of the 14 million deaths of children under five which occur in the developing world each year. The unfortunate reality is that many infant deaths are caused by common ailments, such as diarrhoea and measles, which can be treated at low cost. Diarrhoea, which is the biggest single cause of infant mortality, is simply treated by Oral Rehydration Therapy, consisting of sugar, salt and boiled water.

In 1985, the World Health Organisation (WHO) estimated that 25 per cent of the urban population in the developing world had no access to adequate, safe water and 50 per cent to sanitation. Although the UNDP figures in Table 5.1 (page 106) show more positive statistics, WHO argue that definitions of 'safe access to water' often mean access to a single tap 100 metres away (Cairncross et al., 1990). Although progress was made during the International Drinking Water Supply and Sanitation Decade (1980–90), access in most cities is still far from adequate. Many cities in Africa and Asia have no sewerage system at all. In Calcutta, 3 million people live in 'bustees', plus refugee settlements which lack water and have no refuse or sanitation system, whilst in Bangkok, a third of the population obtains water from vendors (Cairncross et al., 1990). Many poor urban communities can obtain water only through private companies and vendors. In Mexico City, for example, private water trucks visit peripheral squatter settlements to sell water at exorbitant prices, whilst wealthy communities have access to cheap public water supplies. A corrupt system of negotiation between public and private water companies ensures that many needy communities are never connected to the national system (Swyngedouw, 1995). This raises important questions concerning urban power structures and social control over resources, issues which are also relevant in determining who has access to adequate food and nutrition within the city.

Urban malnutrition and access to food

It was estimated in the 1990s that globally, as many as 1 billion people were hungry and undernourished, whilst a further 2 billion risked joining the ranks of the malnourished (Foster, 1992). One factor which is associated with this situation is the rapid increase in the proportion of people living in urban poverty, mainly in squatter settlements and slum tenements (see Chapter 7). Here, high population densities combined with inadequate water and sanitation, as well as insecure employment, hazardous work conditions and environmental pollution from industry, all combine to stretch the coping mechanisms of poor households, as will be examined more fully in Chapter 9.

Malnutrition varies in nature and severity with protein-energy malnutrition affecting 400 million people (World Bank, 1986), of whom two-thirds are children. There are different types of nutritional deficiency, but the most common include anaemia, which is believed to affect half the women in the developing world; Vitamin A deficiency, which is responsible for the blindness of 250,000 children each year; and Vitamin D deficiency, which causes rickets. Dietary and vitamin deficiencies can be treated through governmental food supplements, whereby additives are made to basic staples such as bread. Serious undernutrition, which is measured by a variety of height, weight and age ratios, is often related to deficiences in both urban food production and distribution systems.

The main cause of malnutrition is poverty, but its effects vary according to location. Urban households, many with large numbers of children, have faced declining real wages, the abolition of basic food subsidies, inflation and price rises. Insecure employment and casualised work have led to fluctuating incomes and the inability of many households to provide adequate nutrition and protein. Many urban households lack access to land on which to grow subsistence crops, or are unable to travel to an urban market. Peri-urban agriculture can make a valuable contribution to household nutrition, but frequently squatters and tenants have no access to land (Rakodi, 1985; Drakakis-Smith, 1990, 1991; Simon, 1992). Sanyal (1986) argues that the poorest urban households in Tanzania satisfy their food needs through subsistence agriculture.

Drakakis-Smith (1992) identifies two sources of subsistence production which are common in African cities. Firstly, there are house gardens for those who have access to land, and secondly, there are many illegal plots in marginal areas. Whilst urban subsistence production is important, those without access to land often obtain food through an informal distribution system of vendors. As will be discussed in Chapter 8, petty commodity trading not only provides employment, but is often the only supplier of food to poor communities. Research in one Latin American city found that 64 per cent of slum dwellers could not reach a main market (Pryer and Crook, 1988). Views over the system's efficiency and importance may vary, but there is a growing consensus that the urban informal food system plays a crucial role in food provision for the poor.

In summary, therefore, malnutrition is a direct result of a household's inability to provide basic needs, which is not only related to its size and structure, but conditioned by social and economic forces at the community and national levels. Governmental solutions to urban malnutrition, which include food subsidies and targeting vulnerable groups, are treating the symptoms and not the cause. Other, more far-reaching, policies include the promotion of nutritional education and the development of home gardens. Once again, it is argued that women must be identified as key agents in food and nutrition policies.

Gender and health in the city: maternal and child welfare

Health problems and their causes affect women and men differently (Smyke, 1991). As poverty affects more women than men, it is not surprising that women are more adversely affected by poor health. Thus, Cleves Mosse (1993: 193) states that the 'reasons for ill-health are frequently to be found in the

Table 5.2   Changes in women's life expectancy by region, 1970–1992

Region	Life expectancy (in years)	
	1970	1992
Sub-Saharan Africa	46.3	52.4
Arab States	52.6	63.3
East Asia	64.0	70.6
Southeast Asia and Pacific	53.6	65.5
South Asia	49.0	60.2
Latin America and the Caribbean	63.0	71.0
All developing countries	53.7	62.9
Least developed countries	44.5	52.0
Industrial countries	74.2	79.4

Source: UNDP, 1995

gender role women play, which leads to more sickness and less opportunity for recovery than men'. Østergaard (1992: 110) argues that 'gender, health and development make up a dynamic triad', not least because the basic state of women's health will affect children.

With improved health care, women's life expectancy in the developing world increased from 54 years in 1970 to 63 in 1992 (UNDP, 1995), but as Table 5.2 indicates, this differs greatly between regions. Female mortality still exceeds that for males across life-stages. Current health care practices and social conditions in relation to health all too often act against, rather than in favour of, women's interests.

Women's poor health frequently starts at birth, with greater son-preference and unequal feeding, and high female infant mortality (Table 5.3). A study reported that one in every six deaths of female infants in Pakistan and Bangladesh is due to discriminatory practices (UNDP, 1995). Later on, women suffer hazards from their unequal workload and reproductive activities. Illness associated with pregnancy is seen to be part of a woman's natural burden. Family planning care and advice is frequently unavailable or seen as undesirable (Pearson, 1994).

It is, however, reproduction which takes the greatest toll on woman's health, not least because of the economic value of children and the lack of family planning. The health risks facing a pregnant woman from a poor urban community are indeed great. Maternal mortality is estimated at 500,000 women per annum in the developing world. This amounts to a 1 in 25 risk of death as a result of pregnancy. The risk is even greater in many Asian and African countries where many teenage women are married. Illegal abortions, non-existent care and social notions of birth being 'unclean' all add to this precarious situation. Owing to cultural considerations and notions about women's

Table 5.3   Sex and infant mortality

| Country | Infant mortality (per 1000 live births) | | |
	Females	Males	Female deaths as a percentage of males
Singapore	0.5	0.4	125
Egypt	6.6	5.6	119
Grenada	1.6	1.4	114
Pakistan	9.6	8.6	112
Bangladesh	15.7	14.2	111
Surinam	2.2	2.0	110
Jamaica	1.5	1.4	107
Guatemala	11.3	10.6	107
Honduras	2.9	2.8	104
Algeria	12.8	12.5	102
Peru	5.7	5.6	102

Source: UNDP, 1995

bodies, such as *purdah* in many Muslim countries, women often suffer in silence (Afshar, 1991; Kabeer, 1994). Informal midwives and 'traditional birth attendants' are being seen as key agents in caring for women, but their work goes unrecognised in the formal sense in many cities.

Further links between women and child health conditions indicate the importance of a gendered approach to city health provision. For example, mothers who need to work may not find adequate care for their young children. With increasing demands placed on working mothers, imported alternatives such as formula baby milk and fast foods are promoted by multinationals such as Nestlé in many cities. Misuse of baby milk formula, including the lack of sanitised water, contribute further to infant mortality. The move towards the westernisation of lifestyles via the process of convergence is a major theme here.

Wallace and March (1991) highlight the fact that the prevailing social or legal systems may disadvantage women's health. Hidden attitudes to gender are very important in deciding who benefits from programmes and investments. In India, for example, prostitution is banned under the Prevention of Immorality Act, and this then restricts medical care for prostitutes. AIDS is a further dilemma for many poor mothers in Africa, particularly in relation to transmitting the disease to unborn children, and yet testing is still not available. *Gabriela*, a women's organisation in the Philippines, has pioneered a number of hospitals for women as part of a task force against AIDS.

Despite the immense importance of women's health to city development, most developing countries allocate less than 20 per cent of their health budgets to maternal and child health programmes, with the substantial proportion going

to the latter rather than the former (Herz and Measham, 1987). Instead, programmes and action are being promoted by women at the community level,
often with the help of grassroots organisations. As discussed earlier, gender-
sensitive health care policies are concerned with empowering women so that they
can control their own lives. The concluding section of this chapter analyses
how this is being done, and reports on the influence of urban-based social
movements.

Conclusions: the mobilisation of poor communities

Despite the claim by policy-makers that cities already receive priority in government expenditure and development policy, few of the urban poor have
access to basic needs. As Hardoy *et al.* (1992) argue, public investment in
infrastructure rarely benefits the poor majority, as the settlements in which
they live are located away from schools, hospitals and infrastructure. In relation
to health and education, the marked progress made in the 1960s and 1970s
has largely been undermined by adverse changes in the global economy since
then. As the introductory section of this chapter suggested, the ability of
national governments to alleviate poverty is currently diminishing in all but a
few Southeast Asian nations. Even where service provision has kept up with
demand, the ability of certain specific urban groups to access resources is often
highly unequal.

Increasing criticism has been targeted at misspent revenues on inappropriate
technology and western-style provision of services which only meets the needs
of the few. International agencies such as UNICEF and WHO have adapted
the basic needs ideology to develop the concept of 'basic services', which is
concerned with providing basic welfare provision using appropriate technology
at a grassroots level.

Following the promotion of this grassroots ideology, community organisations funded by NGOs have been identified as better able to empower poor
communities in their fight against poverty. Programmes which enable communities to develop and manage their own solutions have been promoted in
cities across the globe. Whilst critics of this 'self-help' ideology argue that such
approaches enable the state and global institutions to abandon their responsibilities, others recognise that small-scale community developments are making
a real difference to people's lives. In fact, poor urban communities across the
developing world are mobilising for change and making their voices heard.
Castells (1982) notes that cities are the places from where social protests frequently emanate. In this way, traditional power structures are gradually being
challenged, a process that will hopefully lead to social change.

The quality of a community's environment, including its infrastructure and
housing, is generally influenced by those in power. As discussed in the introduction, private landlords, community and church leaders and local politicians
may all have their own agendas in relation to service provision and welfare.

New social movements, many of which stem from women's organisations, are working together to fight for basic needs. From the development of mobile health centres to informal adult education centres, communities are making significant progress. Increasingly, the notion of **community participation** through **civil governance**, involving co-operation between all interest groups, is being seen as a strategy for urban change (Korten, 1996; UNCHS, 1996). Women, in particular, are **empowering** themselves through collective action (Sen and Grown, 1988). For example, an organisation named CIDEM ('The Women's Information and Development Centre) provides a programme for poor urban women in Bolivia, which aims to assist them in gaining control over their own, and their families', health (Cleves Mosse, 1993). CIDEM has set up a 'healthy women' surgery, a pharmacy and a legal aid programme for abused women, and produces educational material for dissemination. It also runs weekly meetings and courses in order to train women to care for themselves. Furthermore, the organisation has promoted national health initiatives and has pressurised the government to co-ordinate public campaigns.

With regard to gender inequality, education is an important tool enabling women to participate in society and improve the standard of living for themselves and their children. An educated population is better able to challenge the political and social system, and fight for justice. Education, however, will have limited success if other basic needs are not met simultaneously. As Cubitt (1995) rightly argues, an educated population unable to find appropriate jobs may lead to social conflict and unrest within the city. Indeed, social protests are not always peaceful and sometimes involve violent confrontation between different ethnic and social groups who are competing for the same resources.

Throughout this chapter, reference has been made to the negative effects of **globalisation** on developing nations. As a concluding point, attention should be drawn to some of the more positive features of current global change, which sometimes go unrecognised. Through the globalisation of the mass media, the importance of human rights and freedom of speech has been promoted throughout the Third World. The global campaign against oppression has, with western NGO support, encouraged marginalised groups to fight for their rights. One facet of this social change relates to freedom of sexuality and 'gay liberation', which until recently has been suppressed in many countries where 'machismo' masculinity is the dominant culture.

The increasing 'Americanisation' and globalisation of Puerto Rico in the Caribbean may well explain the public promotion of gay rights in a recent liberation movement in the capital, San Juan (Plate 5.1, overleaf). Such visible displays of gay identity would still not be possible in other Caribbean nations, where homosexuality is publicly denounced in popular culture, as in Jamaican rap. The San Juan movement is a testimony to the fact that global social change is not always negative, and can assist in the quest for individual freedom and justice within civil society. As 'basic needs' is as much about self-esteem and empowerment as it is about access to shelter, this example offers some hope for the future of the urban poor. If national governments and international

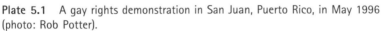

Plate 5.1 A gay rights demonstration in San Juan, Puerto Rico, in May 1996
(photo: Rob Potter).

agencies can harness the creative capacities of urban communities through grassroots development, the next generation might not have to struggle quite so hard to fulfil their most basic of human needs.

Chapter 6

The structure and morphology of cities in developing areas: can we generalise?

Globalisation and city structure in developing areas _____

As noted in the previous chapter, many cities in the developing world have undergone major structural transformations in the last 50 years, not least due to the rapid process of decolonisation which has taken place in Asia, Africa and the Caribbean. As more people are being drawn closer together through new transport and communication links, and changes in commerce and the media, the notion of **global cities** has attracted much geographical attention (Hamnett, 1995; Knox, 1996), as stressed in Chapter 3. Global or world cities, as they are also known (Hall, 1966), are those deemed to be places of intensive interaction which undertake major economic, social and political roles in the accumulation of capital. Global city status is ascribed to those cities which are centres of power and where 'a disproportionate part of the world's most important business is conducted' (Hall, 1966: 7).

As noted in Chapter 3, not all global cities are located in the developed world. Although the term is more likely to include cities in North America and Europe, the prominence of Asian cities and the rising importance of some Latin American centres has added a new dimension to theorising about Third World cities. As cities are dynamic entities, undergoing constant transformation according to prevailing macro-economic and social conditions, it is not surprising that the question of how globalisation affects cities in the developing world has received increasing attention in the last decade (Simon, 1993).

In recent years, the issue of globalisation has attracted much interest in urban studies, through an attempt to analyse the relationship between the structural development of global cities and macro-economic policies (Hamnett, 1995). This new agenda has seen a shift away from theorising which focuses solely on the internal processes affecting city structure, to an approach which aims to enhance our understanding of the ways in which global economic and social processes, discussed in previous chapters, influence city structure and the built environment, and the resulting effect on the social and political organisation of the city.

Much of this recent work, however, has been undertaken in the context of the developed world, and has placed emphasis on western inner city decline, decentralisation and suburbanisation (Sassen, 1991). In global cities such as London, Tokyo and New York, the reconceptualisation has focused on the globalisation of finance and property markets, which has resulted in new waves

115

of urban investment in services and the gentrification of derelict inner city areas. Far less work has attempted to investigate the effects of global economic change on developing world cities, which are also affected by changing patterns of capital accumulation, investment and consumption. In this context, the question of how globalisation is affecting the structure of developing world cities remains more elusive.

Developing world cities are indeed changing, but it is the nature of this change which is often misunderstood. On the one hand, there are common perspectives which argue that as a result of globalisation, particularly in relation to **global time–space compression**, developing world cities are becoming more like their western counterparts, particularly in respect of their consumption patterns, popular culture and the built environment. Evidence to support this argument is apparent in the spread of western consumer goods through 'coca-colonisation', western films, music and multinational enterprises throughout Third World cities across the globe (King, 1991; Sklair, 1991). Such visible outcomes of the **globalisation of culture** have led many protagonists to argue that western cultures are slowly dominating urban lifestyles, and leading to homogenisation across the developing world. At another level, however, it is apparent that western products and popular culture are not readily accessible to all sectors of the population, an argument reviewed in Chapter 3. The key point here is that developing world cities are experiencing a different type of development as a result of their greater internal inequalities. Indeed, the level of social inequality and rapid growth of developing world cities are two factors which distinguish them from advanced capitalist cities.

There are increasing similarities in the central business districts (CBDs) and residential suburbs of many developing world cities due to modern high rise office blocks, impressive multinational headquarters, and the prominence of western retailers (see Plates 6.1 and 6.2). However, the power of western advertising and multinational dominance, which is so readily accepted as a homogenising force, can be questioned by a closer examination of the ways in which globalisation affects the internal structure and social organisation of the city. For example, whilst trends in some developed world cities have been towards decentralisation, suburbanisation and associated inner city decline, there is little evidence to suggest that this is happening on any major scale in the developing world. Indeed, all too frequently, densities are increasing throughout the city. As the present chapter will outline, city centres in many developing regions are still vibrant in relation to their socio-cultural and economic roles (Ward, 1993).

This chapter, therefore, has two main aims. Firstly, it attempts to re-address the relatively neglected theme of the role of developing world cities in the global economy through an examination of the processes which affect city development. It aims to address the issue as to whether developing world cities are becoming more like western cities due to macro-economic forces. For example, how have new transport systems, modern technologies and global capital accumulation affected cities in the developing world? Secondly, through a detailed study of recent work on urban morphology, the chapter investigates

Plate 6.1   Modern and traditional urban forms in the context of Causeway Bay, Hong Kong (photo: Rob Potter).

Plate 6.2   Part of the modern, tourist-oriented sector of San Juan, Puerto Rico (photo: Rob Potter).

whether we can generalise about the structure of Third World cities. Through a critical examination of a number of models which have presented generalisations about African, Asian and Latin American cities, we explore whether a basic model of Third World city structure can be identified.

The chapter begins with a brief account of the social, economic and political factors which are affecting urban morphology and social organisation. Here, attention is paid to understanding the ways in which global processes have influenced city structure. The discussion then deals with the historical dimensions of Third World city structure, through an analysis of **pre-industrial** and **colonial** city forms. A contemporary analysis provides examination of a number of regional city studies and examples, before attempting to draw conclusions. Following on from Chapter 5, the underlying argument focuses on the interactive relationship between urban structure and social process, which has manifested itself in a variety of different urban environments through time.

Cities in the developing world: convergence or divergence? _____

Factors affecting city structure in the developing world

There are two main factors which serve to distinguish developing world cities from advanced capitalist cities. The first is rapid recent growth as a result of rural–urban migration and natural population increase. The second factor is a history of marked social inequality and poverty, where extremes of wealth and poverty are often found in close juxtaposition. Centuries of inequality, often as a result of colonialism, have led to the development of distinct traditional and modern urban modes of production in relation to industry, employment, transport and housing (see Chapters 7 and 8). As earlier chapters have elucidated, developing world cities are increasingly diverse. They have their own unique histories of development which have been influenced by different stages of development, from colonial rule through to their current position in the post-colonial global economy. Thus, we should expect urban morphology to be diverse.

As this chapter will demonstrate, colonialism has played a major role in shaping the internal organisation of cities across the developing world. Socially segregated cities were developed by various European powers from the 1500s onwards, as trading or administrative centres, depending on their size and strategic location. In the post-colonial era, however, cities have been affected by their economic role in respect of their national, regional and global positions as industrial, financial or political centres. Of course, we must not underestimate the importance of macro-economic processes, such as the New International Division of Labour and structural adjustment, in shaping the built environment and the social stratification of cities.

Recently, cities in the Newly Industrialising Countries (NICs) of Asia and Latin America have become focal points for Foreign Direct Investment (FDI), manufacturing branch plants and the headquarters of multinational enterprises, as covered in Chapter 3. Growing consumer markets are demanding modern transport and better urban services. City structure also differs according to a country's ability to undertake the structural transformation, in terms of employment, housing and infrastructure, needed to accommodate increased population. The 1980s debt crisis halted many new plans for urban redevelopment along with the expansion of infrastructure.

An often under-explored facet of urbanisation has been the introduction of modern transport systems which fundamentally alter the location of industry and residential areas, and disrupt traditional social structure (Chokor, 1989). Most cities have a heterogeneous range of formal and informal transport systems which serve different sectors of the population (Hilling, 1996; Simon, 1996). As Simon (1996: 93) argues, transport systems 'need to be understood in terms of the history, culture and functions of each city'. For example, the development of international airports, motorways and mass rapid transit systems (MRTSs) has dramatically altered the social and economic structure of some large cities in recent decades.

In considering the influence of globalisation on urban transport systems, Simon (1996) argues that despite the obvious diversity of cities, the transport systems and management policies being adopted in large cities are converging, as they all generally experienced large-scale motorway and rail developments in recent decades. Increasing car ownership has led to widespread environmental pollution and traffic congestion in Asian and Latin American 'mega cities'. Internationally designed, and in many cases highly advanced, metro systems have become a symbol of modernisation in the NICs of Asia and Latin America. In Singapore, Hong Kong, São Paulo, Mexico City, Caracas and Seoul, to name a few, MRTSs have become a key feature of city expansion. Today, many Southeast Asian cities such Kuala Lumpur and Jakarta are investing huge sums of money in upgrading their current rail systems in preparation for the development of Light Rail Transits (LRTs) which are designed to connect the CBDs with the suburbs. Planning an affordable, sustainable and comprehensive transport system which is accessible to all sectors of the population represents a major challenge to city planners and governments.

Globalisation and developing world cities: convergence or divergence?

A useful initial theoretical framework to employ in exploring the changing role of cities in the developing world is that provided by Armstrong and McGee's (1985) influential work which identifies cities as '**theatres of accumulation**' and '**centres of diffusion**'. As reviewed in Chapter 3 at the urban systems scale, Armstrong and McGee argued that cities play a key role in accumulating capital, international investment, elite populations, modern employment and services. Indeed, the preference of Third World governments and large firms for promoting industrial development in, or close by, large cities has resulted in an important relationship between industrial development and urban structure. Whilst cities provide a plentiful supply of skilled labour and other means of production such as infrastructure, they also house the seats of economic and political power. Cities, however, also play another important role as 'centres of diffusion' with regard to culture, urbanism, modernity, western tastes and consumerism, which promotes further capital accumulation in the city.

This trend is also part of a pattern of '**convergence**' in consumption brought about by elite groups, partly related to the introduction of rapid transit systems, the growth of suburban developments and western lifestyles, whereby developing world cities are becoming more 'westernised' (Harvey, 1989; Potter, 1990, 1993b, 1997). Yet, in respect of many other factors such as demography, social structure, housing quality and productive activities, cities in the developing world are becoming increasingly **divergent**. This argument provides an intra-urban extension to that presented at the urban systems level in Chapter 3.

Thus, an important characteristic of urban structure, both within and between countries in the developing world, is its diversity. On the one hand, urbanisation has resulted in the development of 'mega cities' which have undergone dramatic changes in recent decades, but on the other, smaller cities have also undergone transformations, albeit of a less dramatic nature (Hardoy and Satterthwaite, 1986; Lowder, 1991; Grant, 1995; Gilbert, 1996). It is important to note that a very significant proportion of the rural population in the developing world has migrated to relatively small cities of under 2 million, rather than to the huge '10 million' cities which have captured the attention of the media. Indeed, there are striking contrasts in the scale and extent of city development between continents. As a result, a number of authors have questioned whether a single model of city structure can be widely applied to cities in different parts of the world.

The growth and structure of developing world cities ⸻⸻⸻

Historical development and the pre-industrial city

It has frequently been argued that 'in origin Third World cities and towns – both pre-colonial and colonial – are quintessentially pre-industrial' (Dickenson *et al.*, 1996: 219; Potter, 1985). The concept of the **pre-industrial city** as a distinct urban form was explored by Sjoberg (1960) through a considered analysis of a number of common characteristics shared by developing world cities before industrialisation. In his model, shown in Figure 6.1, Sjoberg identified three main aspects of land use which distinguished the pre-industrial city from industrial cities in developed nations. Firstly, the urban centre tends to be pre-eminent over the periphery, particularly in the spatial distribution of different social classes, as the elites reside in, or near to, the central core, which housed important religious monuments, administrative centres and symbols of prestige. Here, a noticeable characteristic of the pre-industrial city, as distinguished from North American and European cities, is the absence of a Central Business District (CBD). Instead, social patterns in the pre-industrial city are explained by cultural values which identify residence in the historic core as high status. A declining social gradient of residence from the core relegated the

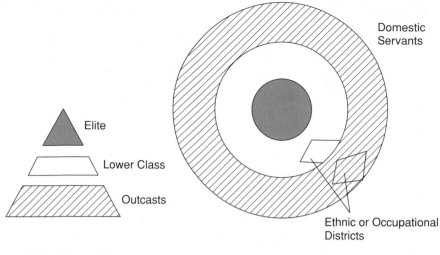

Class Pyramid   **Residential Patterns**

Figure 6.1   The structure of the pre-industrial city (adapted from Langton, 1975).

more disadvantaged groups to peripheral settlements, a pattern which is maintained in many developing world cities today.

Secondly, these broad social areas were further differentiated into distinct quarters or precincts according to occupational status by guild, or to family and ethnic ties. Sjoberg's third characteristic relates to the lack of functional differentiation in land use patterns in the city, where plots were used for multiple functions such as housing and work. Underlying this pattern of mixed land use was the lack of modern transport which predicated not only the use of residences for work, but also explained the elites' desire to live near to the heart of city life. Other common attributes of the pre-industrial city included the prominence of a walled structure and narrow streets which preserved a distinct social pattern, characteristics which are still visible in many cities built before industrialisation and motorised transport. Kano and Ibadan provide good examples of cities which still display many pre-industrial characteristics, but which are increasingly being brought under pressure from modern development (Plate 6.3, overleaf). In both examples, major roads now pierce the traditional city core in order to create modern access for peripheral communities.

In an informed discussion on Third World city structure, Chokor (1989) also maintains that developing world cities are inherently structured around their pre-industrial foundations, with traditional residential communities surrounding a city core, although these have expanded under colonial and post-industrial development. As a result of modern transport and industrial development, most cities now have congested, decaying cores, but they nevertheless maintain a distinct cultural and social structure which is linked to the built environment. As Chokor (1989: 318) argues, 'undoubted unity, social harmony and bonds which exist between physical form and people is thus a key characteristic of traditional city settings'. Examples of this traditional integrative

Plate 6.3 A view across part of the old city of Kano, Nigeria (photo: Rob Potter).

social design can be seen in most developing world cities, whether in the building of walled homes separated by small pathways to give privacy in traditional Islamic and Indian cities such as Hyderabad, the communal courtyards of western Nigeria or the importance of the 'yard' in the Caribbean. As previously discussed, agents of change in the form of western-style planning, particularly in the building of roads and motorways, threaten traditional community structures and ways of life which are essential to the economic and cultural survival of large sectors of the city.

In assessing the overall applicability of the pre-industrial city model to present-day Third World cities, we can see that elements of a pre-industrial land-use structure still exist. It is important to note, however, that Sjoberg's work was based on relatively very small cities of less than 100,000 residents. In this context, Dwyer (1975) has argued that the pre-industrial social pattern is still observable in many small developing world cities, particularly those in Asia and Africa. But other authors have noted that the pre-industrial concept is fast becoming less typical in the rapidly growing 'mega cities' of Latin America.

In many developing world cities, the desire to reside near the city centre is relatively strong, although suburbanisation has increased in recent decades. Amato (1970, 1971), in a series of studies based on Bogotá, Quito, Lima and Santiago, observed that the elite groups have deserted the original city core and moved to one or more sectors of the outer urban fringe. As this chapter will discuss further, trends of suburbanisation by the elites have also been identified in Caracas by Morris (1978) and in Port of Spain by Conway (1981), Potter (1993b) and Potter and O'Flaherty (1995). This process is highly reminiscent of changes which have taken place in western cities. Of course, the fact that developing world cities have now been subject to industrialisation, even if the process has not been uniform, adds a further dimension to Sjoberg's model. In summary, therefore, it is apparent that Third World cities do share some of the common attributes expressed in the pre-industrial model of urban city structure. There is, however, a need for more precise explanations of structure which take account of regional diversity and historical development, a task which has been undertaken by other authors (Brunn and Williams, 1983).

Perspectives on the colonial city

Griffin and Ford (1980) have argued that a contrast exists between thinking about the internal structure of developing world cities and generalised descriptive models of the pre-industrial city and models of modern western cities. It is often presumed that pre-industrial cities will eventually become like western cities, but this line of thought ignores both the distinctive nature of the processes affecting Third World cities, and the diversity of cultural representation in the built environment. One important factor relates to the historical role undertaken by cities in different parts of the globe. In particular, a major criticism of Sjoberg's pre-industrial model is that it fails to differentiate between the diverse colonial experiences of developing nations which comprised distinct social structures (Simon, 1989a).

Horvath (1969, 1972) classified the **colonial city** as a distinct urban form resulting from the domination of an indigenous civilisation by colonial settlers. European-style churches, palaces and administrative buildings fashioned on the prevailing architectural style of the time are visible in most cities. In particular, it is the unique cultural contact between two civilisations which King (1976) identifies as pivotal. According to Horvath (1969: 76), 'the colonial city is the political, military, economic, religious, social and intellectual entrepôt between the colonisers and the colonised'.

Colonial cities were often developed for their commercial functions through the mixing of European urban forms with indigenous tradition, religious and cultural practices. Some colonial cities have been identified as **dual cities** following the development of a new European city next to the old indigenous city, as in New and Old Delhi. The processes through which colonial cities developed have been referred to as **dependent urbanisation**, as cities were reliant on industrialisation located in the metropolitan country (Castells, 1977). The structure of the colonial city is also rather place specific, because it was determined not only by the urban ideology of the colonisers but also by the role of individual cities as agents of imperialism. Some cities were purely administrative centres, whilst others were transport nodes or active in the creation of social surplus product. The structure of the colonial city also differed according to the culture of the colonisers. In the African context, Simon (1989a) states that Portuguese colonial cities such as Luanda and Maputo were different from the British-controlled cities of Lusaka and Harare in respect of their architecture and social organisation.

In respect of internal structure, the colonial settlers retained their position through the development of culturally and ethnically distinct quarters. As a result, colonial cities often maintained interracial social distance which was exemplified in extreme variations in housing quality, the provision of social amenities and access to employment. Elite housing areas for the settlers were often spacious, paved and well serviced, in comparison to the high-density settlements of indigenous populations. Juxtaposed modes of production and employment, consisting of the traditional and modern, remain visible in post-colonial cities today. In fact, it has been suggested by Horvath (1969) and McGee (1971) that cities in Africa and Asia are more likely to have remained traditional than cities in Latin America. As a result of this diversity, a number of authors have attempted to develop regionally specific models of internal city structure.

The African city

Sub-Saharan Africa is still the least urbanised world region, although rates of urbanisation have increased markedly in the last 40 years. On average, African cities are smaller than those of Latin America and Asia, often consisting of between one and two million residents. It has been argued by a number of authors that there is a lack of conceptual studies which attempt to explain the structure and form of cities in an African context (Simon, 1989a). Much theorising presumes that the African city conforms closely to the pre-industrial model, an assumption which underplays the post-colonial development of African cities. Although McGee (1971) argued that smaller African cities are likely to provide the best examples of colonial city structure, it is important to understand the distinct development of African cities under the various versions of European settlement.

An attempt to develop a distinct model of the **African city** was undertaken by the United Nations in 1973. As highlighted in Figure 6.2, the model was based on the existence of an indigenous core and the organisation of different ethnic groups through density gradients in a pattern which ascribed low-density land use to the administrative and housing requirements of the colonial elites, and high density to indigenous populations. In Southern and East African cities, segregation was maintained through strict legislation, whereas divisions were less formal in West African cities, due to the smaller numbers of settlers involved.

The most extreme form of social segregation and economic division was found in South Africa's **apartheid cities**, where the concept of 'separate development' led to the forced removal of the black population from the inner city to townships located on the periphery. Elsewhere in the city, access to urban amenities and employment was segregated according to race and wealth. In the post-apartheid period, attempts to restructure the city and redistribute wealth pose a major challenge for the millennium (Smith, 1992, 1995).

Criticisms of the United Nations' model centre around its failure to recognise the post-colonial transformations which have affected African cities. Post-colonial African cities can be characterised by the mixing of various modes of

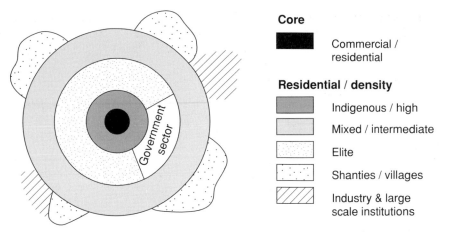

Figur 6.2 A model of the African city (adapted from United Nations, 1973).

economic production and housing, such as traditional and modern, informal and formal (see Chapters 7 and 8), which have resulted in diverse structures (O'Connor, 1983). In this context, however, Simon (1989a) questions whether it is still useful to talk about the 'colonial' or 'post-colonial' city. Instead, he suggests that we might turn our attention to examining the influence of macro-economic forces, such as structural adjustment and neo-liberal trade arrangements, on the future development of African cities.

The South Asian and Southeast Asian City

A great diversity of traditional and modern city forms can be recognised in the great cities of Asia. In particular, there appears to be a distinct difference in the fundamental organisation and structure of the South Asian city compared to cities in Southeast Asia. In South Asia, cities such as Delhi, Calcutta, Bombay and Madras were traditional colonial cities built for administrative purposes, or as ports. Although there is no comprehensive model of the South Asian city, it is argued that they are hybrids of a specific colonial model (Brunn and Williams, 1983; Lowder, 1986), as shown in Figure 6.3 (overleaf). Often built as a port location, South Asian cities frequently contained a walled fort which was surrounded by open space for security. They developed western-styled CBDs, around which the European residences were located. Whilst Europeans lived in well-serviced spacious neighbourhoods, the indigenous population lived in ill-planned streets and high-density housing away from the European population. Although there were some intermediate locations of mixed race, it has been argued that this colonial structure still underlies many modern cities in India.

In colonial times, urban areas in Southeast Asia were essentially developed as trading cities with central cores consisting of local bazaars and high-density commercial uses. Although there were marked contrasts between 'western' and 'non-western' sectors of the city, it is argued that Southeast Asian cities have a more mixed and less uniform structure in terms of land use. In an early model

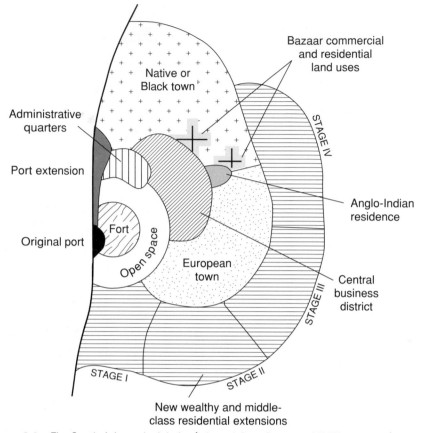

Figure 6.3   The South Asian colonial city (adapted from Brunn and Williams, 1983).

of the Southeast Asian city, McGee (1967) paid greater attention to identifying the various ethnic groups that had been involved in restructuring the
colonial city (see Figure 6.4). Although there is a western-styled CBD, distinct
commercial segments have been developed according to whether the entrepreneurs are Chinese, Indian or European. Although there are zones of squatter
settlements along the city's periphery, new industrial zones and agriculture
are also located in this semi-rural periphery known as the '**desakota**'. McGee
and Greenberg (1992) have examined the way in which these extended metropolitan regions are absorbing village settlements and blurring rural–urban
contrasts.

Also of great importance is the mixed zoning of residential areas between
modern and traditional types of housing and social structure. For example, in
Jakarta, spontaneous as well as traditional villages known as '**kampungs**' occur
throughout the city (Krausse, 1978). Kampungs represent a key component of
Indonesian development, and similar traditional structures can be seen in other
regions. Jakarta's urban ecology broadly conforms to this model, with mixed
inner-city land use and a distinct 'Chinatown' in the west near the port. As in

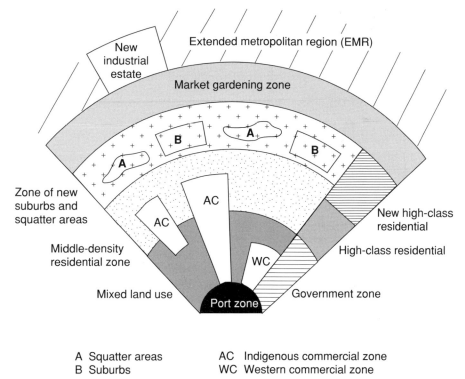

A Squatter areas   AC Indigenous commercial zone
B Suburbs   WC Western commercial zone

Figure 6.4   The Southeast Asian city (adapted from McGee, 1967).

other developing world cities in Africa and Latin America, the elites are starting to suburbanise. This process has been made easier by the development of
mass rapid transit systems in many cities.

   In recent times, given the economic prosperity of the region, Southeast
Asian cities have been the principal centres of growth and planning for social
and economic development (Dwyer, 1972, 1995). Singapore, for example, is
now classified as an important **global city**. Southeast Asian cities have been
seen as catalysts for change, and they have benefited from urban-centred development policies. Despite rapid technological change, however, the majority of
the population still live and work in the non-western areas. This is illustrated
by the existence of large areas of indigenous buildings throughout cities, many
of which still lack modern infrastructure. Whether the 'economic miracle' will
bring about major structural change for the majority of the population remains
to be seen. In essence, therefore, the modern Southeast Asian city is still based
on a dichotomous form of development, but one which is less characterised by
distinct sectional zoning than is the case in the Latin American NICs.

The Latin American city

As noted in Chapter 1, Latin America has undergone rapid urbanisation and
industrialisation in the post-war period (Gilbert, 1990). In 1930 most of the

Latin American population lived in rural areas, but by 1980 over half the population was urban. By the millennium, around one-third of the Latin American population will live in cities of over one million in size, a factor which clearly highlights the rapid nature of rural–urban migration in recent decades. Owing to this unprecedented growth and the effort to achieve modern industrial development in most countries, questions over whether Latin American cities are following the same path as western cities have frequently been raised.

Primacy, the prominence of mega cities and rapid growth from rural–urban migration are all characteristic of Latin American cities. Although there are elements of their development which compare with western cities, global economic processes and national decision making have influenced their internal city structure in a distinctive way. Ward (1993) argues that Latin American cities are inherently different from western cities for a number of reasons. Firstly, at the macro level, Latin American cities have a different demographic structure as a result of post-war rural–urban migration. Secondly, the relationship between macro-economic processes and urban structure has resulted in a different set of outcomes. In particular, Latin American cities have not experienced the same cycle of investment/disinvestment that most western cities have. Through the adoption of industrialisation via import substitution in the 1940s and 1950s, to export-oriented production in more recent decades, the higher level of industrialisation in Latin American cities is also a distinct feature. Furthermore, the rapid social polarisation which has been identified as a major effect of globalisation in western cities has always been inherent in the social geography of developing world cities.

As noted earlier, Latin American cities may be regarded as both 'theatres of accumulation and centres of diffusion' (Armstrong and McGee, 1985) which are currently displaying convergence in respect of the wealth and the consumption patterns of elites, and divergence in terms of social inequality and production possibilities. There has been little evidence of gentrification or large-scale urban redevelopment at the micro scale. Although there have been many plans to redevelop Latin American inner cities, processes of recentralisation were largely halted by economic austerity in the 1980s which curtailed public spending and private investment. It is uncertain whether these plans will resurface if Latin American nations prosper in the new millennium.

Latin American cities are also seen to possess vestigial characteristics of colonial pre-industrial cities. In large conurbations, such as Bogotá or São Paulo, this traditional structure has been superimposed by modern industrialisation, modern transport, the influx of millions of new dwellers and the land-use demands of the elites and emerging middle classes. In an attempt to combine elements of the pre-industrial model with modern developments, Griffin and Ford (1980) devised a model of Latin American city structure in which 'traditional elements of Latin American culture have been merged with the modernising processes altering them'. The original model, shown in Figure 6.5, was based on the Mexican border city of Tijuana, and demonstrates several principal features.

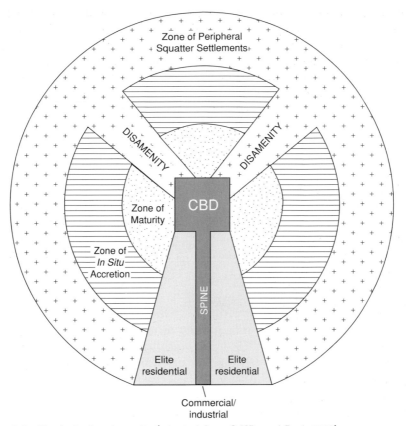

Figure 6.5   The Latin American city (adapted from Griffin and Ford, 1980).

According to Griffin and Ford, the dominant characteristic of the Latin American city is a prominent **commercial spine/sector** which extends out from the CBD, and which houses the city's most important economic, social and cultural amenities, and a substantial proportion of high-income and well-serviced residences (Plate 6.4, overleaf). Residential areas and facilities such as theatres, hotels, restaurants, prestigious offices, private hospitals, museums and leisure facilities are located on, or near to, a '**tree-lined boulevard**', which Griffin and Ford argue is present in nearly all Latin American cities. As shown in Figure 6.5, the elite residential sector is a wedge in the Hoytian sense and combines western-style amenities with a Latin American desire for centrality. Increasingly, the elites have begun to decentralise outwards to western-style suburbs (Plate 6.5, overleaf).

Away from the spine, there is a series of concentric zones with socio-economic characteristics that are almost opposite to those of western cities. Here, socio-economic levels and housing quality decrease with distance from the city centre. Three distinctive zones can be identified: a **zone of maturity**, a **zone of in-situ accretion**, and a **zone of peripheral squatter settlements**. The zone of maturity refers to residential areas which were areas of elite residence, perhaps

Plate 6.4   Part of the arterial core of Tijuana, Mexico, replete with global advertising (photo: Rob Potter).

Plate 6.5   Part of a high-income residential sub-division in Caracas, Venezuela (photo: Rob Potter).

in colonial times, which have filtered down to middle-income groups when elites moved elsewhere, or **'zones of gradual improvement'** where self-built housing has gradually been improved and consolidated over time. These residential areas are also likely to be better serviced than other city districts, with regards to sanitation, transport, electricity and amenities such as street lighting. In many Latin American cities, old colonial residences around the **zocalo** (colonial plaza) have suffered from a lack of investment and upkeep as the relatively wealthy have moved out along the spine, and such dwellings have become rented tenements (Gilbert and Ward, 1982; Gilbert, 1990).

Unlike the relative stability of the zone of maturity, the zone of in-situ accretion is in a constant state of change as residents move in and out according to lifecycle and status. This zone is not uniform, and it comprises a great variety of housing types of a modest quality, often next to commerce and informal economic activity. Some districts are moving towards maturity, whilst others are the recipients of new developments, often self-built or government-maintained. Provision of services such as water and electricity can vary from street to street. Although Griffin and Ford argued that this zone would gradually improve over time, it is unlikely that dramatic improvements have been made in these areas since the 1980s due to economic austerity.

On the periphery of the city are zones of squatter settlements which are characteristic of most developing world cities. As will be discussed in the case of Mexico City, some squatter settlements are so large that they could constitute cities in their own right. In addition to these main zones, Latin American cities also contain sectors of **disamenity**. These are areas of the city which have not been consolidated over time, and are unlikely to be upgraded or improved. Instead, they remain areas of slums and rented tenements such as the **'favelas'** of Rio de Janeiro, where new arrivals to the city are often housed. Major industrial and environmentally polluting activities are also likely to be found here.

With the exception of the commercial spine, Griffin and Ford argue that these concentric zones are essentially pre-industrial, suggesting that if the elites were subtracted, Latin American cities would indeed be pre-industrial. Criticisms of this model have centred on the fact that in many large Latin American cities, elites have begun to suburbanise at a rapid rate due to increasing pollution, crime levels, high-density land use and traffic congestion (Ward, 1990; Gilbert and Varley, 1991). This outward movement of the elites, however, is likely to be restricted by the level of transportation and motorway access to the CBD. Unlike western cities, however, selected suburbanisation has not led to the demise of the inner city which, although suffering from decay in some cases, still retains its overall vibrancy and importance. Ward (1993) has observed that the growth of the urban informal sector is partly responsible for keeping Latin American city centres alive.

In La Paz, Bolivia, a city which is dramatically affected by altitude and terrain, the elites have gradually moved westwards to enjoy a better climate and lower altitude. New shopping malls, property developments and leisure facilities have accompanied this move, although the city centre is still the commercial and cultural heart. In the last decade, a new low-income district of La Paz has sprung up to the east, thus dividing the city into two distinct sections. 'El Alto' is a rapidly expanding low-income settlement which has been built by low-income workers on the eastern plateau around the International Airport. As the name suggests, the settlement is built at an altitude of 4000 metres, giving rise to extreme weather conditions and altitude sickness, in contrast to the western elite suburbs which lie at 2300 metres. Workers from El Alto commute into downtown La Paz, using a variety of formal and informal modes of transport. As a result of this expansion, La Paz is essentially composed of two cities which are markedly different in terms of physical factors such as altitude and climate, and socio-economic factors such as wealth and residential quality. This example further serves to highlight how different cities react to socio-economic change.

The following section critically examines the example of Mexico City, not only in respect of Griffin and Ford's work but also in relation to the effects of recent macro-economic conditions (see Ward, 1990). This examination of Mexico City is based on observations and fieldwork in the 1980s, and also draws on recent research by other authors (Eckstein, 1988; Ward, 1990).

Latin American city structure: the case of Mexico City

Mexico City, perhaps Latin America's premier example of a primary city, is the largest city in the world, with a current population of around 19 million which is forecast to grow to around 30 million in 2010. Mexico City is also the economic and political heart of the country, a factor which is partly attributed to its centralised political system. It is also the world's largest labour market, where around 40 per cent of the population work in the city's informal sector. In addition, 43 per cent of the nation's capital is invested here, and in 1984 it had one-third of the nation's manufacturing and commercial jobs (Dickenson *et al.*, 1996).

In the 1980s, Mexico was adversely affected by rapid rural–urban migration, structural adjustment and declining oil revenues (Eckstein, 1988). As a result, industrial development, employment and urban service provision have all been restricted, resulting in increasing poverty for a large proportion of the population. Despite some optimism concerning increasing economic prosperity in the 1990s, modern Mexico City serves as a good example of skewed urban development, displaying the poverty of the masses and the extreme wealth of the few.

Mexico City occupies the same site as the prominent Aztec city of Tenochitlan, which was built in the early 1300s on the lake of Texaco at an altitude of 2238 metres above sea level. Under Aztec rule, its advanced structure comprised a series of compact rectangular islands and pyramids, which were separated by navigable canals. When the Spanish arrived in 1519, they found a city of over 100,000 inhabitants, with a complex urban social structure. The great city of Tenochitlan was conquered and destroyed by the

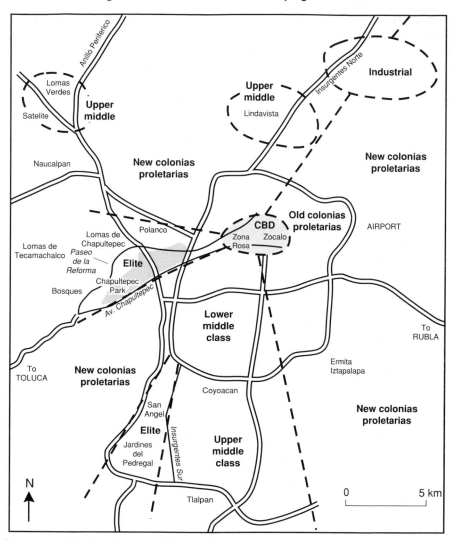

Figure 6.6   Principal features of Mexico City.

Spanish invasion in 1521. A new colonial capital was built on the ruins of the Aztec city, and was renamed Mexico. The colonial city was based on a rectangular grid with a focus on the central Zocalo which contained a large plaza, the national palace and cathedrals, all of which remained prominent after Mexico's independence in 1821.

Throughout history, the Zocalo has remained the historical and administrative centre of the city (see Figure 6.6). In the 1900s, with a population of 345,000, the city's growth largely accrued around the Zocalo, with elite residences being built westwards near Chapultepec Park, and to the north of the city (Ward, 1990). In the 1920s, developments began to extend southwards, and in the 1930s, when the population reached one million, many of the elites built residences in Lomas de Chapultepec, along the tree-lined boulevard of the Avenida de la Reforma.

Since then, and as a result of rapid population growth, the city has constantly expanded its borders to incorporate many old neighbouring municipalities. After the Second World War, the city expanded in all directions, but particularly to the east, incorporating the dry bed of Lake Texaco. This unstable eastern district, which was mainly occupied by lower-income settlements, was the main area of devastation in the 1985 earthquake.

In the 1980s, the city covered over 1000 square km, housing 20 per cent of the national population in 0.05 per cent of the country's territory. The CBD has expanded westwards for over a mile along the Paseo de la Reforma which is known locally as the 'Zona Rosa' (pink zone). This linear CBD houses the city's main commercial offices and skyscrapers, hotels, theatres, exclusive retail outlets and cosmopolitan restaurants. It serves as a vibrant and dynamic centre for the wealthy.

In a pattern which resembles other Latin American cities, the historical Zocalo has gradually filtered down the socio-economic scale, although it still retains an important religious and administrative function. Today, many colonial buildings have been subdivided for rented tenements (*vecinidades*) and low-income requirements. It is, however, still a culturally and economically important centre for lower-income communities, with its informal markets, fiestas and religious services. In attempting to draw upon Griffin and Ford's ideas on Latin American city structure, therefore, Mexico City's spatial development broadly conforms to some generalisations, but as noted by other authors (Ward, 1990, 1993), there are other processes affecting the city which are not considered in their analysis.

As stated above, Mexico City does have a commercial spine which extends from the CBD westwards along the tree-lined boulevard of Paseo de la Reforma, where many of the city's elite residences of Lomas de Chapultepec and western-styled amenities are located. In the 1970s, this sector was extended farther westwards to incorporate the pleasant, if rather hilly elite suburban developments of Bosques de las Lomas complete with American-style shopping malls, golf and health clubs, and private security patrols. Areas such as Polanco, which consists of houses and apartments close to the CBD, appear to be filtering down slightly to upper-middle class residents. Other elite residential areas include San Angel and Jardines de Pedregal, which offer old colonial-style residences in the peripheral southwest of the city. The congestion, pollution and crime of central Mexico City are leading many elites to move to greener peripheries. Western expansion, however, is limited by the mountain range, although there has been some movement outside the city to Puebla, Cuernavaca and Querétaro. In 1981, the government invested in a plan to electrify the rail link between Querétaro and Mexico City to enable commuters to move out of the city, but this was halted by the ensuing debt crisis. The extent to which elite groups will continue to suburbanise remains to be seen. One major obstacle is the lack of services, transport and economic activity to meet their demands outside the city.

With some exceptions, the quality of Mexico City's housing stock is extremely poor outside the western–southern section and presents a complex social structure. In 1980, 30 per cent of homes lacked an internal water supply and 40 per cent were without sewerage facilities (Ward, 1986). As Figure 6.6 (page 133) highlights, there are zones of maturity around the central city. Established areas of squatter settlements known as *colonias populares* (people's neighbourhoods) constructed in the 1950s have now become part of the formal city structure, and are reasonably well serviced. Elsewhere, new *colonias populares* are growing on the city's periphery, without access to urban services and employment. Generally, lower-income neighbourhoods, *colonias proletarias*, are situated to the north and east of the city where the airport and major industry are located.

Areas of slums and shack developments, known as *ciudades perdidas* (lost cities), also constitute part of these developments. The growing middle class is located mainly around the southwestern wedge, such as Coyocan, along the Insurgentes Sur motorway where status increases with distance from the city. In the 1970s, satellite developments were built to the west and north of the city for middle and upper-middle class residents in Satelite, Tlalpan and Lindavista which were linked by El Periferico to the CBD.

Satelite was the site for Mexico's first large shopping mall development and leisure complex, but since the 1980s the town has been encroached upon by *colonias populares* and is now part of the city. Many areas of the city that were once peripheral and rural in character have since been absorbed by the city's expansion. Some *colonias populares* on the city's periphery are the size of many developing world cities. For example, Naucalpan in the northwest has grown from 30,000 in 1950 to 408,000 in 1970 and to 1 million in the 1980s. Since the mid-1980s, more rural areas to the north and east of the city have been consolidated by low-income dwellers at an alarming rate.

From this example, it can be seen that although Latin American elites are showing some signs of convergence in their residential and consumption patterns, Latin American cities are evolving in a culturally specific form. Furthermore, the scale of inequality and the sheer extent of urban deprivation in places like Mexico City have not been seen elsewhere. Since the 1980s, increasing environmental degradation, poverty and congestion in Mexico City have pointed to decentralisation as a planning policy, but whether this can really be achieved remains debatable.

The structure of developing world cities: final comments

This chapter has focused on the changing internal structure and social organisation of cities in the developing world. The key theme has been the recognition that developing world cities, despite some evidence of westernisation, have reacted in diverse ways to the forces of globalisation and macro-economic change. Whilst the applicability of various models of urban morphology to cities in Africa, Asia and Latin America is highly debatable, these models do highlight a number of common similarities which serve to enhance our understanding of traditional city structure.

Since the 1960s, rapid rural–urban migration has changed the character of cities in the developing world. Although it can be argued that some cities maintain various characteristics of their pre-industrial form, it is this recent growth which has had the most profound effect on city structure. In the post-modern era, many cities are experiencing some convergence in terms of architectural style, modern CBDs, leisure facilities, transport developments and popular culture, but this change is expressed in many different ways according to distinct social and cultural environments. In terms of their social structure, most cities in the developing world are still socially segregated in terms of access to wealth, urban services, employment and adequate shelter. Thus, developing world cities are not following exactly the same path as western cities. If nothing else, the existence of marked social inequalities, which are visible in the massive squatter settlements surrounding such cities, make them distinct urban forms in their own right.

Chapter 7

Housing and shelter in Third World cities: rags and riches

Introduction

It is sobering to reflect that around 20 per cent of the world's total population is thought to be lacking decent housing. With regard to the residents of the developing world, it is estimated that at least one fifth, and perhaps over half, live in substandard housing, and this is in situations where national governments are either unwilling, or unable, to make housing available of a higher standard.

Housing is of vital importance to social welfare and to the development process as a whole. This was indicated by the fact that the United Nations declared 1987 as the International Year of Shelter for the Homeless (IYSH). The importance of providing adequate shelter had earlier been recognised by the United Nations Conference on Human Settlements (Habitat) held in 1976. The Habitat II or 'City Summit' conference was held in Istanbul in June 1996 (Berghall, 1995; Okpala, 1996; UNCHS, 1996). Food, clothing and shelter, and the gainful employment to provide these, are vital prerequisites for all our lives, and are essential ingredients for any basic needs approach to development.

In the majority of major cities in the developing world, more than one million people live in illegally or informally developed settlements, with little or no piped water, sanitation or services (McAuslan, 1985: 6). The residents are frequently unable to afford even the smallest or cheapest professionally constructed, legal house with basic amenities. The majority of extant houses have been **self-built** or **autoconstructed** in that their residents have taken responsibility for organising the design and building of their own houses. In the early 1960s, Abrams (1964: 1) bemoaned the fact that despite progress in the fields of industry, education and the sciences, the provision of simple refuge affording privacy and protection against the elements was still beyond the reach of the majority of the world's population. In 1990/91 it was estimated that 9.47 million people or 60 per cent of the population of Mexico City lived in self-help housing. At about the same time, 1.67 million or 61 per cent of the residents of Caracas were to be found in such shelter.

Three major themes are stressed in the present chapter. Firstly, the issue of housing is ultimately inseparable from the wider issues of inequality, structural poverty and social welfare. By virtue of this there is great commonality between all developing countries. Secondly, the housing problem is as much about land ownership and access to land as it is about housing itself, and what-ever the local circumstances, the role of the state is fundamental, as stressed in Chapter 4. However, as a third point, it is necessary to recognise that the precise configuration of housing type, land market and tenure system is very varied and highly country-specific. Each of these themes is elaborated in this chapter.

Spontaneous self–help housing in the developing world: definitions and processes

The housing problem in the developing world has really emerged since the early 1940s. As Abrams (1964) stressed, there are at least three distinct types of poor urban dweller in developing world cities. First, there are the **homeless** and the **street sleepers** (Plate 7.1). In Calcutta in the early 1960s, it was estimated that more than 6)0,000 slept on the streets, whilst in Bombay, one in every 66 was homeless and a further 77,000 lived under stairways, on landings and the like (Abrams, 1964).

Secondly, a large group are to be found **renting accommodation in slums and tenements**. It is recognised today that there are many renters in self-help settlements (see Gilbert, 1983; Gilbert and Varley, 1991; Kumar, 1996), along with the existence of squatter landlords (see Gilbert, 1983; Lee-Smith, 1990; Potter 1995). Thirdly, there are the **squatters** and **occupants of shanty towns**. In Cairo, as an example, the severe housing shortage has led to a number of novel responses. The old city, or medina, has become a vast area of tenement slums. Perhaps more surprising are the tomb cities, or cities of the dead, which are to be found located on the eastern edge of the city. Here, the tomb houses

Plate 7.1   A rickshaw driver sleeps in the streets of Delhi, India (photo: Rob Potter).

built for caretakers or for relatives visiting graves are now occupied by the poor as permanent homes.

Another novel response in Cairo is living on the rooftops of apartments, for as long as the structures placed on existing roofs are not constructed of permanent materials, they are permitted. It is estimated that in the region of half a million people live in such rooftop dwellings in the city (Brunn and Williams, 1983; El-Shakhs, 1971; Abu-Lughod, 1971). The growth of population in low-income settlements is frequently running at between 12 and 15 per cent per annum (Turner, 1967; Dwyer, 1975). Frequently, shanty towns and squatter settlements account for at least 20–30 per cent of the total urban population, but on occasion the proportion is far higher, as for Bogotá (60 per cent), Casablanca (70 per cent) and Addis Ababa (90 per cent).

It is the squatter settlement or the shanty town that is the most ubiquitous sign of rapid urban development in the developing world. Such settlements are also referred to by means of a wide variety of other descriptions, among them **spontaneous**, **informal**, **uncontrolled**, **makeshift**, **irregular**, **unplanned**, **illegal**, **self-help**, **marginal** and **peripheral settlements**. These adjectives are qualified by means of an almost bewildering array of local names such as the *barriadas* and *pueblo jovenes* of Peru, the *favelas* of Brazil, *colonias proletarias* of Mexico, *gecekondu* of Turkey, *bustees* of India, *bidonvilles* of Algeria and Morocco, *gourbivilles* of Tunisia, and many others besides.

It should be stressed that spontaneous self-help responses have occurred in developed countries, for example in the United States, Germany and France in times of economic hardship (Harms, 1982, 1992). Indeed, as Abrams (1964) reminds us, before the capitalist mode of production, self-help universally represented the predominant form of housing provision. Dwellers have also constructed self-help dwellings in the United Kingdom in the form of plotland developments, riverside homes on the Thames and elsewhere, prefabricated units, and beach huts on the south coast (see Hardy and Ward, 1984). In December 1995 there were reports of the development of extensive squatter-type dwellings by Moroccan migrants in Madrid.

The wide variety of labels used to describe such settlements points to an important characteristic, namely their extreme diversity with regard to formation, building materials, physical character and their inhabitants. The terms **squatter** or **illegal settlement** are frequently used, but are potentially misleading. Squatter settlements are those where individuals have settled without legal title to land, or alternatively without planning permission. A good example of the latter is provided by the *barrios clandestinos* of Oporto in Portugal. Squatter settlements are frequently located on government- or church-owned land. But illegality is not always a characteristic. Many low-income homes are owned, the plots having been subdivided and sold. Similarly, some homes and/or the land on which they are sited are rented. Such **rentyards** are quite common in many Caribbean and Latin American towns and cities (Clarke and Ward, 1976; Ward, 1976; Potter, 1985).

Yet another common characteristic is that areas are **makeshift settlements** or **shanties**, being constructed from whatever materials are available to hand.

Plate 7.2   Low-income housing in Port-of-Spain, Trinidad (photo: Rob Potter).

Basic shelters made from packing cases and fish barrels, as well as cardboard cartons and even newspapers, have been described in the Moonlight City area of West Kingston, Jamaica, by Clarke (1975). More typically, recycled scraps of wood and corrugated iron may be employed, together with printers' drums for both outside walls and roofs. Other materials that are frequently employed in construction include flattened tin cans, straw matting and sacking.

Such settlements may also be makeshift in the sense that they have none of the basic urban services such as water, electricity or sewerage when they are initially developed. But even here caution must be exercised, for with time such services are often acquired, and brick-built houses may come to predominate in a formerly makeshift area. Certainly, the basic but frequently overlooked distinction between squatter settlements characterised by their illegality on the one hand, and shanties identified by virtue of their poor physical fabric on the other, must be fully appreciated and borne in mind (Plate 7.2).

Yet other settlements are characterised by their unplanned, irregular and informal nature, or by their origin in mass land **invasions**. Such haphazard or speedy development is epitomised by the description **spontaneous**. All of these terms are highly appropriate in certain situations, but are potentially misleading in others. Thus, whilst many low-income settlements are unplanned in the professional planning and architectural senses, many are the outcome of much careful forethought on the part of their residents, especially those involving organised land invasions, which may occur at the suggestion of opposition politicians (see Gilbert, 1981; Potter, 1994). However, in Africa, Asia, and the Middle East, the development of low-income housing is typically a much more gradual process, being based on slow infiltration and individual initiative. Such developments are, therefore, based on the very antithesis of spontaneity.

Similar concern may be expressed concerning the lack of universal applicability of descriptions such as **peripheral** and **marginal settlements,** whether used in a strictly geographical or an economic sense. Finally, although the terms **self-help** and **autoconstruction** are useful in signifying that the building of such dwellings is not normally undertaken by professionals, it would be highly erroneous if the impression were given that such houses are built entirely by their present or previous occupants. Indeed, often the assistance of friends and family is enlisted, supplemented by artisans, as in the *coup d'main* system of St Lucia (Phillip, 1989; Potter, 1994).

In short, the very character of these settlements necessitates a catholic definition, and the adoption of a single description merely for the sake of convenience. The expressions **low-income**, **informal** or **self-help housing** are preferred here, so that the range of housing responses of the poor can be seen together, without implying a misleading homogeneity among inhabitants. However, the critical point is that these settlements are very diverse, varying with respect to basic attributes such as materials, tenure, legality, speed of occupation, size, location, the origin and nature of residents and their aspirations.

Spontaneous self-help housing: locational aspects _____

The urban poor in the developing world cannot afford houses that are professionally or formally surveyed, built and serviced. Where rental property is available, rents are frequently exorbitantly high. Thus, poor citizens often construct peri-urban structures on land which has not previously been used for building purposes. Typical sites include small vacant plots within the old walled section of the city, as in Manila (Dwyer, 1975). Another typical site is on steep hillsides, as exemplified in Caracas, Hong Kong and Rio de Janeiro. Land which is swampy or subject to flooding offers further opportunities, as shown by the example of Singapore. Similarly, land adjacent to railways is also often occupied, as in the case of Kuala Lumpur. Recently reclaimed areas are also frequently colonised.

In the mid-1970s the British geographer Denis Dwyer presented an extremely useful model of the location of spontaneous settlements. Starting from the observation that low-income settlements are to be found throughout the city, it was stressed that increasingly, the largest were peripheral. Secondly, the city centre with its nexus of urban job opportunities, both formal and informal, can be regarded as the principal attracting force for spontaneous settlements. Dwyer (1975) noted that as the urban area expands, so the ideal squatter location starts to move out centrifugally by the normal processes of invasion and succession. Other prime locational factors for spontaneous settlements are good water supplies and tolerable relief. However, with growth, spontaneous settlements may be displaced outward towards difficult topographical sites and those which lack water supplies, as shown in Figure 7.1A (overleaf). Additionally, agricultural spontaneous settlements on the urban fringe

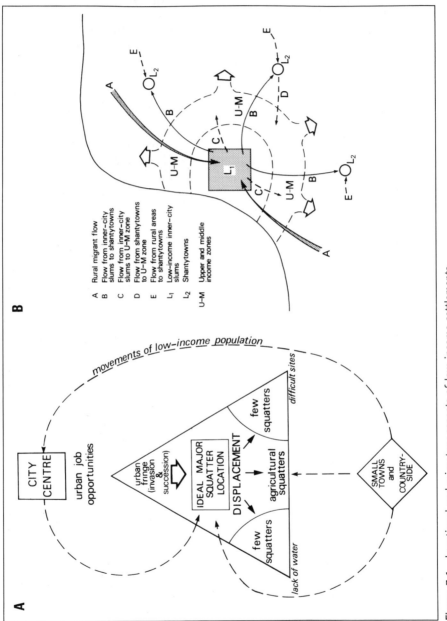

A

movements of low-income population

CITY
CENTRE

urban job
opportunities

urban
fringe
(invasion
&
succession)

IDEAL MAJOR
SQUATTER
LOCATION

DISPLACEMENT

difficult sites

few
squatters

agricultural
squatters

few
squatters

lack of water

SMALL
TOWNS
and
COUNTRY-
SIDE

B

A   Rural migrant flow
B   Flow from inner-city
    slums to shantytowns
C   Flow from inner-city
    slums to U-M zone
D   Flow from shantytowns
    to U-M zone
E   Flow from rural areas
    to shantytowns
L₁ Low-income inner-city
    slums
L₂ Shantytowns
U-M Upper and middle
    income zones

Figure 7.1   Locational and migratory aspects of low-income settlements.

may be incorporated into the urban mosaic. The model stresses the association of low-income settlements with difficult and poor sites, and the environmental implications of this are explored in Chapter 9.

The location model is also useful in stressing that there are diverse migratory paths that are followed by low-income groups in order to reach such settlements. The popularly held view that low-income settlements are inhabited by migrants who have recently moved directly to them from small towns and the countryside is indeed frequently true in many African cities. However, for most cities in Asia, Latin America and the Caribbean, the evidence suggests that spontaneous settlements do not act as the settling or reception areas for new arrivals from the rural zones. More typically, new urban arrivals go directly to the city centre in order to be close to the nexus of job opportunities, as shown on the outer right-hand loop of Figure 7.1A. Thus, inner city tenement slums often act as the reception areas for new migrants, who only later move to spontaneous settlements, frequently after establishing themselves in the urban labour market. Increasingly, this distinction is seen as quite basic and suggests yet another dimension of diversity in that squatters are not the poorest of the poor.

A case study: Montego Bay, Jamaica

In 1972, Alan Eyre of the University of the West Indies published an excellent study of the shanty towns of Montego Bay, the second largest settlement in Jamaica. In the early 1970s, Montego Bay had a population of around 50,000. Eyre noted that strong spatial variations in income and wealth characterised the city. Further, it was suggested that Montego Bay could be split into six broad zones: (i) the central business district; (ii) the tourist sector; (iii) the free-port commercial area; (iv) an outwardly expanding arc of middle-income residences; (v) an outwardly expanding arc of upper-income residences, and finally (vi) a group of shanty towns. In total, 10 shanty towns were identified, mainly located some four to six kilometres from the urban centre, and between them housing in excess of 20,000 people, thereby representing some 40 per cent of the total population.

Eyre identified five main flows of population underlying the structure of the city, as shown in Figure 7.1B. These are denoted from A to E in the diagram. The relative magnitude of these flows substantiates the earlier discussion and illustrates the salience of Dwyer's location typology. The predominant flow (A) was of rural migrants directly into the low-income inner-city slums. Most of those involved in this movement were reported to be single and poor and looking for their first job.

The second most significant flow (B) was shown to be that from the city centre slums to the shanty towns. These individuals were seen as wanting to escape from the high rents and other negative aspects which characterise inner-city life, such as crowding and crime. Eyre characterised this group as those with families, and perhaps with some capital too. The other flows involve noticeably fewer people. They include the movement of the upwardly mobile from inner-city slum areas to the middle- and upper-class residential zones (flow C) and an even smaller flow from shanty towns to the middle-class ring (flow D). Finally, the flow of rural migrants direct to the shanty towns (flow E) is shown to be of only very minor significance.

Eyre's study stressed that the majority of shanty town dwellers are not rootless rural migrants, but urbanites of long standing. The research showed that the average household head had lived in the Montego Bay urban area for a period of 11 years, including time spent in the squatter areas. Further, more than three-quarters of the population of the 10 shanty towns had been born within the city limits. The rural migrant flow was primarily directed towards the inner city, where rows of slum tenements specifically cater for such transients. It was noted that a central location is more convenient for job seeking, since peri-urban location incurs travel expenses (Eyre, 1972).

In a recent follow-up study, Eyre (1997) has pointed to the fact that after several decades of turning a blind eye to squatters in Montego Bay, as of early 1994 the Jamaican authorities have started to try and clear long-established squatters from areas such as the land at the end of the runway at the international airport. Eyre notes that this change in policy mainly reflects the fact that such land is now wanted for commercial purposes within the context of deregulation and neo-liberalism. This is a general point to which we shall return later in this chapter.

From spontaneous self-help housing to 'oppressive housing' _____

In the 1950s and 1960s, self-help housing was generally viewed with alarm and pessimism, representing a problem which had to be cleared and replaced by regular housing (see Lloyd, 1979: 53–57). Such negative and pejorative views were reflected in the writings of the American anthropologist, Oscar Lewis. Lewis worked in Mexico, India and Puerto Rico and he argued that the poor were locked into an inescapable '**culture of poverty**' (Lewis, 1959, 1966). Lewis argued that wherever poor groups are found they show traits such as apathy, fatalism, tendencies towards immediate gratification and social disorganisation.

These simple deterministic ideas came to be attacked from a number of quarters (see Lloyd, 1979; Drakakis-Smith, 1981; Gilbert and Gugler, 1992). In particular, it was suggested that the culture of poverty is convenient for the wealthy and the powerful, insofar as it suggests that 'poverty is the poor's own fault' (Gilbert and Gugler, 1992: 84). It is views such as these that are reflected in popular descriptions of squatter settlements and shanty towns as 'urban cancers', 'festering sores', 'urban fungi' and the like. During the period when top-down modernisation was unquestioningly accepted as the route to development, it was inevitable that poor housing would be viewed erroneously as the problem, and not the direct outcome and reflection of poverty.

The first response to these concerns can be referred to as the '**state as provider**'. In a number of early instances, in clearing squatter settlements, the state went on to build houses for the poor. As an example, in the period after 1954 through to the mid–1980s, the Hong Kong government was responsible for the construction of over 400,000 new domestic residences. These now accommodate over 2 million people, or approximately 44.5 per cent of the total population of the city. The scheme, which stands as one of the largest public housing programmes in the Third World, was basically designed to settle squatters and not as a means for assisting low-income families *per se*. Dwyer (1975), Drakakis-Smith (1979) and Yeh (1990) argue that a strong motive on the part of government was to free sites occupied by squatters for more lucrative permanent development.

The programme commenced in 1954, having been precipitated by a major fire in the Kowloon area which had left 53,000 squatters homeless in the previous year (Yeh and Fong, 1984). It was decided to embark on a large-scale programme of high-density accommodation using a very basic housing design. The early resettlement estates provided only 2.2 square metres of usable floor-space per adult resident (Yeh, 1990). Thus, the standard unit was a six- to seven-storey H-shaped block. All facilities were provided on a communal basis, including water taps, bathrooms and toilets, which were located in the central crosspiece of the blocks. Some 240 blocks of this type were built before 1964, and these house half a million people.

It is now generally accepted that the standards adopted in these early schemes were far too low. Accordingly, a major effort to improve the quality of resettlement housing was made during the period 1964–73, during which time the building programme was extended and upgraded. After 1964, newly built blocks were larger, at least 16 storeys high, and the individual housing units were more spacious, having their own toilets, water taps and kitchens. On the negative side, however, there was an increased tendency for these new estates to be located on the urban periphery, away from the main employment opportunities.

Since 1973, squatter resettlement has been retained as the principal plank of housing policy, but emphasis has also been placed on the improvement of the early resettlement estates. Although Hong Kong's housing programme has not been without its critics, especially in relation to the very high residential densities employed, it stands as one of the most ambitious resettlement schemes ever attempted in the Third World. However, it is generally agreed that the government's motive was the production of scarce land for urban development rather than a response to squatter needs, or wider welfare considerations (Keung, 1985; Yeh, 1990).

Another example of a government housing scheme instigated with the specific aim of eliminating squatter settlements is provided by Caracas, Venezuela. Caracas grew rapidly following the development of the oil industry in the early twentieth century. In 1950 it had a population of 0.6 million, and by 1981 this had grown to just over 2 million. The city is in a narrow valley, and self-help housing areas or *barrios* had become a feature of its expansion. The housing programme was carried out by the Banco Obrero, the principal housing agency, at the instigation of the dictator Jimenez. This crash programme started in 1954, and between then and 1958 some 97 high-rise developments or 'superblocks' were built, 85 in Caracas and 12 in the nearby port of La Guaira. In all, over 16,000 apartments were provided in these superblocks. Most are 15-storey blocks which are luxurious by comparison with the rancho areas.

The headlong speed of the rehousing programme is attributed to President Jimenez's desire to make an overt show of his power and his apparent concern for the social welfare of the people. It is now agreed that the outcome was little short of disastrous. Many ranchos were bulldozed with little notice and the residents transferred to the superblocks, which had been extremely hastily designed. For example, a major fault was the interior location of the stairwells. Jimenez was overthrown in 1958, by which time the superblocks were in near chaos (Dwyer, 1975; Drakakis-Smith, 1981; Potter, 1985). For example, some 4000 families had invaded empty flats, whilst rent arrears had reached staggering proportions. Squatter shacks had been built close to the blocks, notably on green spaces provided between them. There had been little or no physical maintenance of the blocks, but far more damaging was the fact that adequate social facilities had never been provided.

The largest group of buildings, located to the east of the CBD and known as the *Urbanization 23 de Enero*, had come to be dominated by political factions. With the overthrow of Jimenez, the programme was suspended, and subsequently, evaluation surveys were carried out. These served to emphasise that the low incomes of the residents made it impossible for them to pay the rents and maintenance charges without subletting. Such overcrowding further exacerbated the social problems of high-density living. A major recommendation was the need to build up community involvement by means of tenants' associations. In 1961, a large-scale social work programme commenced in the 23 de Enero area. It has been argued that Caracas affords perhaps the most infamous example of an ill-planned and misdirected mass public housing programme in a Third World city.

There are, of course, many further examples where the state has endeavoured to provide houses for low-income groups, such as Barbados, where Potter (1989a, 1992a) and Watson and Potter (1993, 1997) show how the local self-help house has been ignored. This happened first when barrack-style and row housing schemes were introduced, and later when concrete starter houses were used as the central plank of state housing upgrading in the 1980s. In the meantime, the policy relevance of the local vernacular architecture has been almost totally neglected (see also Potter, 1986a; Potter and Conway, 1997).

Most authors are agreed about the lessons that are to be drawn from such examples. In a nutshell, apart from a few wealthy city states, most Third World governments cannot afford high-technology, high-rise monumental responses to their housing problems. But more significantly, nor can the mass of poor people in these countries. The headlong rush into high-technology, western-inspired housing schemes seems singularly inappropriate and incongruous when viewed both environmentally and socially. Inevitably, over time, such **oppressive housing** tends to filter up to the middle classes. This has led to the general argument that if developing countries are to improve housing, the government must not build houses. In short the state should not be the provider.

From spontaneous self-help to aided self-help

As a result of failures such as those chronicled in the last section, new research began to challenge conventional wisdoms. In a paper presenting an overall theory of slums, Stokes (1962) drew a clear distinction between what he regarded as successful and unsuccessful poor communities, referring to these respectively as **slums of hope** and **slums of despair**, a terminology which has stuck (but see Eckstein, 1990). Charles Abrams' (1964) book was also influential, for he stressed that urban land costs were soaring, thereby pricing the poor out of the market. But he believed that there was sufficient land available if only it could be appropriated by public sector intervention. He noted in particular that in the conditions of a housing shortage, the bulldozing of houses represents a curious policy.

As Conway (1985) summarises, the perspective concerning self-help housing swung from negative to positive in the late 1960s to the early 1970s. Gradually there was a change in perspective concerning informal sector housing and employment. It was argued that the poor were not indolent, dishonest and disorganised, but generally, in fact, quite the reverse. The major change in attitudes was to be precipitated by the experiences of two academics-cum-architectural/planning practitioners who were working in Peru in the late 1960s. The first was William Mangin, an American anthropologist, and the second John Turner, a British architect-planner. Both Turner and Mangin advocated **self-help housing** as a positive force in developing world housing provision. One of the most important papers was written by Mangin (1967), the title of which conveys the essence of the overall argument presented by the two authors: 'Latin American squatter settlements: a problem and a solution'.

In his work, Mangin described most of the then dominant views on low-income residents as myths. They were not disorganised, a drain on the urban economy, dominated by criminals and radicals, nor were they made up of a single homogeneous social group. Rather, Mangin stressed that most squatters were in employment, were socially stable and had been residing in the city for a considerable period. Illegal occupancy of land gave them the opportunity to avoid paying high rents, and at the same time allowed them to build their own homes at their own pace.

In like manner, Turner worked for over eight years in Peru, and for a considerable proportion of that time he was involved with self-builders in various *barriadas*. His partly autobiographical account (Turner, 1982: 99–103) is very informative in this connection. His overall attitude is clearly summarised in one concise quotation:

> Like the people themselves, we saw their settlements not as slums but as building sites. We shared their hopes and found the pity and despair of the occasional visits from elitist professionals and politicians quite comic and wholly absurd. (Turner, 1982: 101)

Turner argued that all that had to be done to assist self-builders was to approve rough sketch plans, and to distribute small amounts of cash in appropriate stages. Turner observed that the economies of self-help were founded upon 'the capacity and freedom of individuals and small groups to make their own decisions, *more* than on their capacity to do manual work' (Turner, 1982: 102). As a consequence, Turner articulated the Churchillian cry, 'never before did so many do so much with so little' (Turner, 1982: 102).

Table 7.1   Turner's typology of low-income housing groups in developing countries

Housing group	Housing priority		
	Proximity to the urban centre	Permanent ownership	Modern standard of amenity
Very low-income 'Bridgeheader'	✓	✗	✗
Low-income 'Consolidator'	–	✓	–
Middle-income 'Status seeker'	✗	✗	✓

✓ key housing priority; ✗ very low priority; – intermediate priority

The most positive message promulgated by Turner was that if left to themselves, low-income settlements improve gradually but progressively over time. Thus, houses that were originally constructed from straw matting later acquired walls, services and paved streets. In the terminology of Stokes, they were clearly slums of hope, characterised by *in situ* improvement and the generalised upward social mobility of their populations (see also Turner, 1963, 1967, 1968a, 1968b, 1969, 1972, 1976, 1982, 1983, 1985, 1988, 1990). By such means, the **use value** of the property, reflecting its utility as a basic shelter, is slowly transformed into higher **exchange values**, reflecting the market valuation of the dwelling.

Turner's most basic ideas concerning housing improvement can be rendered as a simple typology of low-income housing groups in developing cities (see Turner, 1968a). This is summarised as Table 7.1. Put in simple terms, Turner argued that in making residential choices all individuals are influenced by three major groups of factors. First, there is the need to live in close proximity to the inner-urban ring, so as to be near available job opportunities. Secondly, there is the desire for permanent ownership and/or security of tenure. Thirdly, with rising real incomes, the quest for modern standards of amenity applies to some. But the relative importance of these three desires varies for different sets of low-income residents, giving three major sub-groups, referred to as **bridgeheaders**, **consolidators**, and **middle-income status seekers**. These three groups reflect the recurrent theme of the socio-economic diversity existing among low-income housing groups.

The very low-income **bridgeheader** is normally a recent migrant who is seeking to become established in the city. His or her priority will be to find a job and thus to reside close to the urban core, whereas the need for permanency, ownership and high standards of amenity will be minimal. In contrast, the low-income **consolidator** has already gained a firm foothold in the urban economy. It is more than likely that his or her income will be rising and that a high priority will be placed on attaining better standards of amenity and

Table 7.2 The characteristics of informal and formal sector housing

Housing characteristics	Informal		Formal (consolidated)
	Early stages	Consolidating	
Responsibility for construction	users/occupiers	users/occupiers	suppliers (state agency or private developer)
Legal status	illegal	illegal/legal	legal
Conformity with standards	none	some	conforms to standards
Level of infrastructure	absent, improvised	improvised and incomplete	usually complete
Income level of inhabitants	low	range of income levels	range of income levels, usually higher
Tenure	*de facto* ownership	variety, increase in renting	variety
Professional involvement	no involvement	sometimes limited involvement (NGOs)	dominant
Physical nature	precarious	temporary/permanent	permanent
Activities in dwelling	shelter	mixed: income generation	mixed (may be restricted by regulations)
Use/exchange values	use value	increasing commodification	exchange value
Mode of production	artisanal	mixed	manufactured/industrial
Labour for construction	household	household/paid labour	contractor

Source: adapted from Kellett, 1995, p. 30

ownership. Turner's model thereby parallels the division of settlements and their inhabitants into the categories of those based on **hope** (consolidation) and **despair** (bridgeheading). As shown in Table 7.1, the model also allows for a third group of fully upwardly-mobile individuals, who are described as **middle-income status seekers**, for whom modern standards of amenity become the ultimate goal, whilst proximity to inner-ring areas and permanent ownership are factors of little or no consequence. Kellett (1995) has usefully summarised the characteristics of informal sector housing, from its early stages, via consolidating housing, through to formal and fully consolidated housing, as shown in Table 7.2.

The major policy implication of Turner's work was, of course, that governments are best advised to help the poor to help themselves by facilitating **spontaneous self-help**, and by fostering and facilitating **Aided Self-Help** or what has become known as 'ASH'. There are three principal forms of ASH: (i) the upgrading of existing squatter housing, (ii) the provision of site and service schemes, and (iii) core housing schemes, where the shell of a house is provided on a site. These approaches will be fully elaborated, with examples subsequently in this chapter.

The critique of self-help housing

The process of consolidation of low-income housing undoubtedly occurs if the circumstances are right. Two conditions are generally seen as being essential prerequisites to spontaneous housing consolidation: firstly, a sufficiently high income among residents, and secondly, security of tenure. Without these, neither the means nor the desire for substantial housing improvement and upgrading are likely to exist.

It is salient that much of the attitude-changing work described above was carried out in the Latin American context, where despite the existence of wide income differentials, economic conditions have generally been more prosperous than elsewhere in the less developed realm, a point stressed in Chapter 1. Thus, in Latin American towns and cities, incomes on average have been high enough to stimulate the consolidation process. It is also tempting to note that the pioneering work of Mangin and Turner was carried out in the desert conditions of Lima, Peru, where undeveloped land was freely available on the urban periphery.

In Africa and Asia and much of the Caribbean, conditions are by and large poorer. Thus, other things being equal, there is less chance that self-help imperatives will be able to alleviate housing problems. This is an important cautionary note, which is certainly borne out when contemplating the housing problems of, say, Calcutta and Bombay. It also serves to remind us that despite the many commonalities, the developing world cannot be treated as a homogeneous entity, a point also emphasised in Chapter 1.

Turner's ideas are just as noteworthy and perhaps as contentious too, in suggesting that members of the public know what is best for them. Clearly, Turner's overall thesis challenges elitist forms of planning and architectural practice. Similarly, ideas concerning the acceptability of individual and community self-help dovetail with those advocating appropriate or intermediate technology (Dunn, 1978) and indigenous, grassroots development. Self-help housing, involving the original building of homes, their improvement and the lobbying of local politicians, planners and other state functionaries, can all be seen as primary forms of public participation in the planning process in the cities of the developing world. Participation on a representative basis is a time-consuming and demanding process (Fagence, 1977; Conyers, 1982; Potter, 1985; Desai, 1995a). The sort of approach that is required can be seen from the work of Andrews and Phillips (1970), who by means of consultation exercises involving sample surveys, examined the attitudes of inhabitants of selected *barriadas* in Lima. It was established that the local population was most dissatisfied with the provision of property titles, the location of medical facilities and the enhancement of utilities such as street paving, lighting and water.

Although the promotion of self-help appears to have democratic appeal, and has certainly helped to change both governmental, public and indeed international opinions regarding low-income settlements, many have observed that it potentially embodies other less laudable and progressive connotations. The principal advantages of self-help include the provision of alternative housing, labour rather than capital intensity, overtones of intermediate technology, institutional acceptance and powerful backing by organisations such as the World Bank and the 1976 Habitat International Conference.

However, a large number of analysts have maintained that the acceptance of an explicitly self-help based housing policy can be used as an excuse for inactivity and neglect by government. In other words, whilst it may well be best for governments not to build houses, they should be doing all manner of other things, ranging from providing technical expertise on site and in the form of self-help manuals, to appropriate employment and land policies. In respect of land, the need is for government to acquire urban land, by either **land banking** or **land pooling**. In the former, land is acquired in advance of need; in the latter, a public authority acquires land with many plots and owners, consolidates the area, and eventually reallocates demarcated and serviced plots to its former owners in proportion to their original holdings (McAuslan, 1985). As stressed in several places in this text, land access has to be seen as a major factor in Third World housing. The middle classes compete for land, and when this is added to the demands of the elite groups in society, the poor's situation is further disadvantaged. All too frequently land is held by the state, the church, the military, and by the middle and upper classes for speculative purposes.

These arguments led to what has become a fundamental critique of self-help housing developed by Marxist-oriented writers, who maintain that Turner's thesis and the policies flowing from it serve to maintain the *status quo* of monopoly capitalism (see Ward, 1982; Mathéy, 1992a, 1992b, 1992c; Marcuse, 1992; Fiori and Ramirez, 1992; Ramirez *et al.*, 1992; Burgess, 1982, 1992; Harms, 1982, 1992; Betancur, 1987; Aldrich and Sandhu, 1995; Potter and Conway, 1997). It can be argued that if taken to extreme, the ideology of self-help romanticises and rationalises mass poverty, and makes light of the lack of access to land and capital of the poor majority (see Burgess, 1977, 1978, 1981, 1982, 1990, 1992; Conway, 1982; Harms, 1982; Potter and Conway, 1997). Burgess (1982) goes as far as dubbing Turner's advocacy of self-help housing 'a curious form of radicalism'. Harms (1982: 18) argues that the people's so-called freedom to build for themselves confuses freedom to act with the need to survive. Even Abrams (1964) observed that the poor in developing countries are doubly exploited: once at work and then again at home as part of the house building and improving process.

It is a short step to argue that self-help reduces the wage bills of commercial enterprises by reducing the capital costs of building and maintaining relatively poor housing conditions. In turn, this leads to the argument that self-help as a comprehensive and generic policy serves to reduce the costs of the social reproduction of labour in the future. It is but another short step to argue that self-help housing is a vital ingredient in maintaining a reserve army of labour. Indeed, there is specific evidence illustrating governments turning a blind eye to, or even encouraging, the development of urban squatter settlements during periods of economic boom and high demand for labour. Main (1990) shows this clearly in the case of Kano, Nigeria, and he also demonstrates that during periods of recession with declining demand for labour, the city authorities have on several occasions embarked on squatter clearance programmes. Equally, there are countless examples of governments taking action to remove squatters when land is required for more lucrative commercial purposes (Potter, 1994; Potter and Conway, 1997).

It is certainly tenable to suggest that this overall interpretation casts some very interesting light on government reactions to illegal squatter settlements. Sometimes this amounts to what appears to be a policy of benign neglect, where the issue is effectively ignored. On yet other occasions or in other contexts, eradication is a principal response. In Manila, for example, in the 1960s, at one stage 2877 shacks were demolished in a two-week period (Dwyer, 1975). But, as stressed earlier, the eradication of even substandard dwellings amounts to a bizarre form of housing policy unless, as a result, the occupants come to acquire more satisfactory shelter. It is tempting to suggest that the state's posture in turning a blind eye in some circumstances reflects the fact that such a strategy affords great power over the groups which are occupying land illegally. Further, the provision of key services and utilities immediately prior to an election is not uncommon. In a similar vein, the question of political patronage crops up once again when, as is common, the residents of squatter settlements argue that they have been encouraged to occupy government-owned land by the opposition political party.

However, despite these concerns, it was perhaps almost inescapable that the positive aspects of low-income housing should come to be used as a principal plank of housing policy in developing world cities. But as with all palliatives, there is the possibility that its very success in the policy arena may serve to prevent examination of the fundamental causes of the problem. These are, of course, structural and relate to income inequality, land ownership patterns and mass poverty. Exactly the same type of argument will be reviewed in the next chapter in relation to the provision of work and employment in the developing world city.

Articulation theory may be used as a framework to summarise these circumstances and arguments. Referred to as 'articulation of the modes of production under capitalist development', rendered in simple terms the argument runs that capitalism seeks to **conserve** traditional pre-capitalist forms wherever these work in its favour, but specifically seeks to eradicate or **dissolve** those which are not directly advantageous (see, for example, McGee, 1979a; Wolpe, 1980; Burgess, 1990, 1992; Ward and Macoloo, 1992; Potter, 1995). Hence, it is posited that the local self-built house is retained (see Plate 7.3) due to its cheap and efficient nature in relation to furthering capitalist production, whilst

Plate 7.3 A self-help builder and children in Barbados, West Indies (photo: Rob Potter).

in other areas of the economy, traditional forms are seen as outmoded and are replaced with modern forms, such as middle-income housing, supermarket-based retailing, and the like.

Aided self-help housing in practice: principles and examples

Slowly in the late 1970s, a new orthodoxy was being adopted by the principal international aid agencies, specifically the World Bank, United Nations and United States Agency for International Development (USAID) (Williams, 1984). In this reappraisal of views, two main ideas were conflated. Firstly, it was accepted that financial resources were too limited to allow centrally funded mass housing schemes. Secondly, the existence of high levels of under- and unemployment among the poor meant that many hours of labour were available, so that leisure time could be converted to 'sweat equity'. These two ideas came together and gave rise to the view of the state as **facilitator** or **enabler**, and not as provider (see Hamdi, 1991).

Thereby, spontaneous self-help gave rise to aided self-help. Beginning in 1972, a substantial proportion of the World Bank's urban lending programme was based on the proposition that site and services and squatter upgrading projects afforded the best path to urban improvement (Conway, 1985). Although site and services take a number of different forms, by the early 1980s these complementary approaches were being employed in 100 different countries (Laquian, 1983; Payne, 1984). The scale of the adoption of the approach was quite remarkable. Williams (1984) shows that the World Bank alone committed US$1200 million per year to aided self-help housing programmes, whilst USAID were putting up US$100 million per annum during the 1970s.

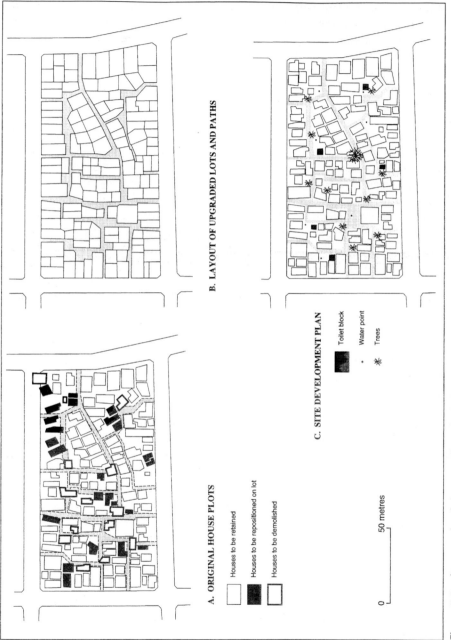

B. LAYOUT OF UPGRADED LOTS AND PATHS

C. SITE DEVELOPMENT PLAN

Toilet block

. Water point

☀ Trees

A. ORIGINAL HOUSE PLOTS

☐ Houses to be retained

■ Houses to be repositioned on lot

☐ Houses to be demolished

0   50 metres

Figure 7.2   An example of the upgrading of a low-income housing area.

Pugh (1995) notes that from 1972 to 1990, the World Bank participated in 116 aided self-help housing projects in some 55 different countries, entailing an average expenditure of US$26 million. As Payne (1984) dryly observes, Bank missions with full portfolios are far more effective in persuading governments of the utility of an approach than even the most articulate of academics.

Upgrading is the simplest and probably the most effective form of aided self-help (Martin, 1983). Upgrading not only preserves the existing low-income housing system, it also retains the economic system used by the poor, along with the local community structure. This is particularly important as it means that child care and employment facilities can be retained. There are important gender issues here, which have only recently been given the attention they deserve (see Moser and Peake, 1987; Chant, 1991, 1996b), for it is all too often women who have made a disproportionate effort to establish and develop homes in low-income subdivisions (Peake, 1997). The alternative to upgrading is relocation in one form or another, with all the disruption and increased transport costs that this is likely to entail.

There are three major components to the upgrading of slums and squatter settlements. The first is the **provision of basic services**, such as potable water, toilets, surface drainage, garbage collection, streets, footpaths, schools and perhaps even community centres. The **rationalisation of the layout and alignment of structures** represents a second objective. Such 'reblocking' inevitably involves the dislocation of existing structures: sometimes as many as 50 per cent of homes have to be either moved on the site or relocated elsewhere (see Figure 7.2). There are salient potential gender, age, community participation and other issues in this connection as well. If by chance the houses which need to be moved off-site belong to elderly, mainly female household heads, the infirm or members of other disadvantaged groups, then a strong case exists for exchanges of lots to be made in the interests of social justice, so that those more able to accommodate to the move should do so. Such a process needs the mobilisation of effective community participation in the planning process. Thirdly, upgrading involves the granting of **security of tenure**, in the form of either outright ownership or the provision of long-term leases. The formal definition of the boundaries of residential plots is as important a part of this overall process as the provision of legal title.

One of the best examples of squatter upgrading is provided by the Tondo Foreshore Development Project in Manila, in the Philippines. This is an area of 180 ha of reclaimed land, which is adjacent to the port waterfront, where 27,500 families were living as squatters. The project commenced in 1976 at an estimated total cost of US$65 million, approximately half of which came from a World Bank loan. The main approach was a normal upgrading exercise, comprising reblocking and the construction of roads and paths. The actual site plan for a part of the area is shown in Figure 7.2. The scheme also involved the introduction of piped water and waterborne sewerage, along with community centres and schools (Laquian, 1983).

Housing loans were made to families in materials, not in cash. Land titles were provided as leaseholds, nominally paid on a monthly basis, but with holders having the option of purchasing titles outright after five years. In all, approximately 12,000 homes were upgraded and 4000 moved to new site and service schemes as a result of the reblocking process. Despite the evident success of the scheme, Crooke (1983) argues that there is evidence that as the Tondo area was rapidly incorporated into the city's formal housing market, so increasing exchange values have meant that wealthier families have moved in, thereby effectively jeopardising the area's role as a low-income residential area. However, despite this tangible expression of some of the Burgess fears, similar schemes have been transferred to the national scale.

An example which further demonstrates some of the potential difficulties that surround Turnerian ideas is provided by the experience of slum upgrading in Bangkok, Thailand. In 1980, the World Bank recorded that some 30 per cent of the city's population lived in high-density, inner-city slum areas, which were grossly deficient in basic urban services (Crooke, 1983). But, on the other hand, as a consequence, land and dwelling rents had tended to be low enough to be affordable to the poor. The majority of these houses are on privately owned lands, where dwellings are normally rented, either directly from the landowner or through intermediaries who rent and sublet.

The existing ownership of slum dwellings continues unchanged, as the expropriation of slum lands would not be politically feasible unless the owners were compensated at full market value. Thus, the programme of upgrading carried out with World Bank assistance has concentrated on infrastructural upgrading alone, providing all-weather walkways, land drainage and water supplies. The Bangkok example shows the problems of upgrading privately owned slums, for if the market value of the properties rises as a result of even infrastructrual improvements alone, the eventual benefits can easily accrue to the landowners and not to the low-income residents. Thus, any rise in property values caused by upgrading will soon be reflected in increased rents. In order to prevent this happening, the Thai authorities made the owners of the upgraded slums undertake to keep rents at their extant levels in real terms, and further to guarantee tenants security of tenure for a full 20 years after the initial upgrading exercise. In practical terms, this exemplifies the need for the state to be involved, especially in land value and access issues, if housing projects are to work.

The other major policies which have gone hand-in-hand with upgrading have been **sites and services** and **core housing**. The former involves the opening up of new land and its subdivision into residential plots which are serviced with utilities. Alan Turner (1980: 268) has commented that 'this term has become a generic title for a wide array of projects and has acquired an almost talismanic quality of being a cure-all for the housing problems of the poor'. There is great variation in such schemes, in terms of both the services provided and whether the initial stages of the dwelling are provided. Skinner and Rodell (1983) attribute the broad concept of 'land-and-utilities' to the practical works of Abrams and Koenigsberger between 1955 and 1963.

The shelter component of such schemes can vary from no shelter at all, that is a strictly defined site and services, to more sophisticated levels of provision. In the former case, the remnants of shanties cleared from elsewhere may be used (see Potter, 1992a, for example). Where the beginnings of the house are provided, this is referred to as the '**core unit**'. Sometimes just the toilet or '**wet core**' is provided on site. More frequently, a fire wall and toilet, plus a kitchen and perhaps even the first room, are provided. The standard of services provided can vary greatly as well. For instance, with regard to water provision, this may be merely by means of a common standpipe for approximately five households, through to water piped directly into each house. Likewise, toilets may vary from pit latrines to flush toilets. Finally, lot size also ranges widely, from pegged-out lots of 45 square metres as in the case of Tondo, to 300 square metres or more.

Many early site and service schemes were located on the urban periphery where land is cheap. By such means, planners are able to influence the directions in which they prefer the future growth of the city to occur. But in some early schemes severe difficulties of disruption occurred due to the distance of such schemes from the workplaces of the residents, and in a number of instances projects had to be abandoned for exactly this reason. Together with access to employment, the provision of security of tenure is fundamental to such projects. Lots may be either sold or leased to residents.

Another important dimension relates to the size and sophistication of the houses which are built on the lots. Rarely do housing authorities control the size of houses directly. Rather, they present model houses as guides. In some instances, this has proved troublesome, for they are often of a high standard, reflecting the middle-class professional aspirations of architects and politicians. At the same time, politicians are unable or unwilling to offer subsidies. In such circumstances, there is the chance that the policy becomes little more than a sick joke.

A good example of a core housing scheme is afforded by the Dandora project in Nairobi, the capital of Kenya. This was a large-scale attempt by the Nairobi City Council and Government of Kenya to provide low-cost housing, with the planned provision of some 6000 serviced plots over a seven-year period. The World Bank, the International Bank for Reconstruction and Development (IBRD) and the International Development Agency all supported the project. Lots of between 100 and 160 square metres with individual water and sewerage connections, access roads, lighting and refuse provision were planned. Wet cores and demonstration houses were provided, and there were three main options, from wet cores at 100 square metres, to wet cores with kitchens, stores and one room at 160 square metres.

It is notable that aided self-help has not been confined to capitalist countries. In Cuba, since 1979, some 60 per cent of all new housing units have been constructed by voluntary 'microbrigades'. These take the form of squads of workers who second themselves from their jobs in order to build housing for their fellow workers (see Hamberg, 1990; Mathéy, 1992b, 1997). In turn these workmates are expected to work overtime in order to maintain production levels in the absence of the microbrigade workers (Hall, 1981b). Recently, the policy has been extended to the upgrading and redevelopment of pre-existing communities within the historic urban core of Havana, under the title 'social microbrigades'.

Plate 7.4   Low- and high-income residential areas cheek-by-jowl in Colaba, Bombay (photo: Janet Carlsson).

Concluding comments

Once the political rhetoric is placed to one side, few would regard efforts to provide basic low-income housing at an appropriate level as being unaccept- able, whether by self-help or indeed any other means. But, on the other hand, any housing policy which merely provides homes without any effort to reduce poverty has to be seen as spurious and unacceptably partial (Plate 7.4).

Hence, it is easy to follow the lead of Alan Turner (1980: 253) in suggesting what he sees as the seven basic needs for the enactment of a balanced housing policy: (i) upgrading existing slum areas; (ii) ensuring an adequate supply of land for new settlements; (iii) providing security of tenure for residents; (iv) provid- ing access to appropriate finance; (v) wider social and economic development, by means of job creation schemes and small business loans; (vi) appropriate technology; and (vii) the passing of legislation in support of these aims. This represents a wide and holistic remit, but it serves to emphasise one of the major themes of this chapter, namely that the housing issue is about far more than dwellings. Hence, even if the state does not build houses, its responsibil- ities in the arenas of housing and land are not obviated. Similarly, self-help housing can only help, if such an approach to housing is accompanied by other policies (Potter and Conway, 1997; Aldrich and Sandhu, 1995). Without progress on a wide front, little is likely to be achieved in tackling the pressing housing problems that face the residents of most Third World cities.

Chapter 8

Employment and work in the developing world city

Introduction

One of the disturbing features of development since 1950 has been the un-precedented growth of population and the urban labour force. The inability of countries to provide sustainable employment for a growing labour force consti-tutes one of the most powerful obstacles to social and economic transformation in the developing city. In the late 1980s, it was estimated by the International Labour Office (ILO) that between 1985 and 2000, a total of 1000 million new jobs would have to be created in order to establish full employment worldwide (ILO, 1988).

As highlighted in Chapter 3, developing world cities attract a large number of job seekers. The global urban population increased from 737 million in 1950 to 2603 million in 1995 (ILO, 1995). As noted in Chapter 1, interna-tional agencies estimate that by 2005, more than half the world's population, that is around 3350 million people, will live in urban areas. It is envisaged that most of this growth will occur in developing countries, accounting for some 660 million new urban dwellers, compared to 87 million in industrialised countries (UNDP, 1993).

Habitat II in 1996 brought together global institutions, national and local governments and NGOs to address the most salient issues facing developing cities, namely escalating poverty and inequality. By focusing on the two major themes of sustainable human settlements and adequate shelter for all, the conference stressed the need for equity in urban policies (Drakakis-Smith, 1995; Jones, 1996). As there are important links between urban production, employment, social equity and poverty alleviation, the ability of cities to pro-vide sustainable work for their growing populations is imperative. As Drakakis-Smith (1996b: 673) argues, 'Cities function by drawing on the skills and labour of their populations: in turn, people are drawn into the city in search of work and opportunities to improve their lives.' It is people, therefore, who must be the focus for policies which attempt to readdress urban vulnerability and poverty.

Rapid urban population growth rates pose particular problems for employ-ment creation in developing cities across the globe, where over 300 million residents live in absolute poverty (UNDP, 1995). Furthermore, global trans-formations in production, trade and finance have placed a strain on tradi-tional coping mechanisms at the international, regional and household scales.

Sustained labour force growth is predicted over the next 25 years, with the greatest increases occurring in Africa and the Middle East. By 2015, sub-Saharan Africa will add annually to the world's labour force more than three times as many new workers as countries from the Organisation of Economic Co-operation and Development (OECD), eastern Europe and the ex-Soviet Union taken together (Turnham, 1993).

As a result, work is one of the key social and economic issues in the contemporary world (Pahl, 1988). The concept of work has been subject to critical debate in a western context, but there is still much confusion about its meaning in the context of the developing world, particularly in the light of recent global transformations. There has been much debate over the concepts of 'work' and 'employment'. In its narrowest definition, the term **employment** is given to work undertaken in return for a wage or other remuneration. Attempts to measure employment rates, however, have frequently included only employment in the formal economy which is accounted for in national statistics, thereby omitting other forms of employment undertaken in the informal economy or the household.

A country's **labour force** consists of those people of working age, often between 16 and 65, who are able to work and are therefore in the **labour market**. Those people who are employed in the labour market are referred to as the **economically active workforce**, although this definition often excludes people who are economically active but not accounted for in the statistics. It is becoming increasingly accepted that 'employment' should refer to all types of productive work which is remunerated in some fashion. **Unemployment** refers to those who are registered as not being economically active and seeking work. However, as this chapter will stress, the relevance of the term 'unemployment' is questionable in many developing societies.

The concept of **work** extends to include all productive or social activities, including those which are unwaged, which make a contribution to the household, community or country. Here, voluntary community activities and household reproduction and subsistence production are all valued. The sphere of work, which incorporates a wide range of unwaged and unpaid activities, is critical to the understanding of the structure and operation of the developing world city. After all, it is cities such as São Paulo, Bombay and Mexico City which are currently experiencing the effects of a rapidly growing labour force. More specifically, the nature of urban work and employment is becoming a critical issue for policy makers, due to escalating poverty, increasing unemployment, and the inability of cities to provide waged employment for their growing workforces. As work and employment opportunities influence personal and national wealth, poverty and status, their consideration is central to the understanding of current social, economic and development issues. Employment patterns have a strong influence on urban development, both in respect of the accumulation and distribution of wealth, and in relation to the structure of the built environment.

As such, the 1990s presented a new set of challenges over and above the problems of recession, falling production and structural adjustment which were experienced during the 1970s and 1980s. Neo-liberalism, the power of multinational corporations, and the development of huge trading blocs in Europe, North and Central America and Asia all presented new obstacles challenging sustainable work opportunities for the majority. Neo-liberalism has promoted 'free trade' as the path to economic development based on the relative success of a few Asian economies. The promotion by World Bank and OECD economists of foreign direct investment into export-oriented industry as a solution to unemployment in many developing countries has caused much concern over the quality and future sustainability of such newly created jobs in the 'South'. Whilst free trade is promoted as bringing prosperity for all, the free market is not free for those who are unable to compete on an equal footing. Furthermore, OECD countries still retain many barriers to trade in areas where developing countries have their greatest advantages. The creation of GATT and new trading blocs such as the European Community and the North American Free Trade Agreement (NAFTA) will have far-reaching effects on the lives of many workers across the developing world. NAFTA promotes free trade between 365 million people in two wealthy nations, the US and Canada, and the much poorer nation of Mexico. In this unequal exchange, it is likely that NAFTA will gradually turn the whole of Mexico into a one large '*Maquiladora*' for producing North American exports (Coote, 1995), whilst at the same time flooding the country with American products. Although NAFTA's ability to create new manufacturing jobs in Mexico is not disputed, its role in promoting equitable development for the Mexican poor is questionable.

Critics who contest the far-reaching effects of globalisation need only witness the increasing exploitation of Third World workers or the rising incidence of child labour as national responses to neo-liberalisation and free trade. As the *World Employment 1995* report states, 'anxieties over the issue of job creation have surfaced against a backdrop of profound change in the global economy' (International Labour Office, 1995: 5). The anxieties mentioned refer to deteriorating employment conditions and widespread underemployment for a majority of the labour force in developing countries. Official unemployment and underemployment in developing nations currently stands at 40 per cent of the potential workforce (UNDP, 1992). As a result, a large proportion of the labour force is seeking alternative forms of employment, outside the boundaries of officially defined work, giving rise to the **informal sector**.

In most of the developing world, with the exception of a few successful Asian economies, formal sector job creation has declined at a time of rapid labour force growth. There has been worldwide concern over growing unemployment since 1970 (ILO, 1970; Turnham *et al.*, 1990; Turnham, 1993). Quite simply, urban labour force growth outstrips job creation. With an estimated 2.3 per cent per annum increase in the working population in developing countries (Turnham, 1993), employment creation cannot keep pace with demand, even in regions where there is increased industrialisation. Furthermore, real wages have fallen whilst the costs of basic needs have increased, particularly in Africa, Latin America and the Caribbean, where employment prospects have worsened since the 1980s with the implementation of structural adjustment policies (Gilbert, 1994; Simon *et al.*, 1995; Cubitt, 1995; Green, 1995; Stewart, 1995).

As a result of declining opportunities in the formal sector, developing cities are becoming characterised by large **informal economies**. Even despite new opportunities created by export-oriented industries, growth rates in manufacturing employment have failed to keep up with urban population growth. But the informal sector also reflects increasing recession and poverty, as domestic industry contracts and multiple jobs become the means to survival. Many countries in sub-Saharan Africa have witnessed the collapse of their modern employment sectors in the 1980s. As noted in Chapter 3, substantial improvements in job creation have only been made in the Asian Newly Industrialising Countries (NICs), but even here the benefits are not spread evenly (Dixon and Drakakis-Smith, 1995). The employment 'success' of East Asian cities has been attributed to the reduction of surplus labour in agriculture by increasing employment opportunities in manufacturing and services. Although, due to successful export-oriented economic growth, in the 1980s formal employment grew at 2.3 per cent per annum in Hong Kong, and 5.9 per cent in Singapore, many urban workers remain underemployed. Whilst manufacturing employment grew by 6 per cent per annum in Singapore, the informal sector was still growing due to high rates of immigration from neighbouring countries.

Elsewhere, large proportions of many countries' labour forces are still employed in agricultural occupations or in the urban informal sector. Agricultural employment still accounts for 70 per cent of the working population in sub-Saharan Africa, and 65 per cent in South Asia, whilst the urban informal sector continues to expand as rural employment opportunities and standards of living decline. Recently, the UNDP has estimated that the informal sector accounts for 80 per cent of employment in Nigeria, 68 per cent in Bombay, and 59 per cent in Latin America (UNDP, 1996). It is further argued that between 1980 and 1993, 82 out of every 100 new jobs created in Latin America were in the informal sector.

Labour absorption by the economy also differs by region. A recent categorisation of developing countries according to their employment structures (Turnham, 1993) highlights East Asia, Thailand, Malaysia, Chile and Indonesia as the most successful countries in terms of employment creation and reducing unemployment. A second category refers to the middle-income countries of Latin America and the Middle East, where workers have moved from rural to urban employment, resulting in the growth of the urban informal sector and limited formal manufacturing growth in certain industries. A third category includes rural low–middle income countries in Asia, for example, China, parts of India and Indonesia, where urban employment is similar to the second group. Africa is relegated to the fourth category, where there is high labour force growth, a large agricultural labour force, poor prospects for urban formal employment and a large informal sector in the major cities. A common trend appears to be the growing importance of the informal sector in providing urban employment in all but the most modern of the developing economies.

It is important to note that whilst employment problems are becoming increasingly connected to processes of globalisation, the influence of national economic and social factors, such as gender, race and class, on individual workers remains important. As discussed in Chapter 5, the ability for developing cities to empower civil society through the process of urban governance depends on co-operation between global institutions and local organisations. Whilst globalisation has increased the demand for Third World labour, and in places has created new jobs, it has also limited the effectiveness of traditional policies to create employment in domestic industry. Whilst the benefits are largely accrued outside the developing world, members of the global workforce have witnessed the negative consequences of deregulation, growing inequality and social polarisation. As a result, the challenge to provide secure work becomes more complex.

The remainder of this chapter presents an analysis of the structure of work and labour force participation in an urban context. In it, particular attention is paid to the gender division of labour and the effects of globalisation on women's work. Drawing upon recent examples, the aim is to illustrate how city dwellers adapt their work practices to economic, political and social change at a range of scales, from the international to the local. In the next section, generalisations concerning the nature of employment patterns and urban labour force segregation in the developing world are presented. Consideration is given to both unremunerated labour as well as waged work, unemployment and underemployment, the gender division of labour, and the conceptual issues underpinning the nature of work. Subsequently, a further section examines theoretical perspectives on the structure of urban economies in the developing world, before focusing on the nature of work outside the formal economy, namely informal employment and unwaged work.

Following this, the relationship between global processes and local action provides a structure for analysing workers' responses to current social and economic trends. By means of discussion of the New International Division of Labour (NIDL), subcontraction, household production and child labour, the effects of neo-liberal economic strategies, dependency and structural adjustment on work patterns in the city are analysed. In conclusion, consideration is given to current political thinking on employment and the implications of policy action in the city.

Work, employment and the urban labour market

An overview of changing urban labour markets

A major consequence of rapid urbanisation has been the growth of the urban labour force in both the formal and informal urban economies. For the majority of developing countries, urban unemployment is twice as high as the rural rate, ranging from 10 to 20 per cent (Todaro, 1995). Unemployment is particularly high amongst the 15–25 age group. Although the inability of countries to provide jobs for a growing labour force has been a global concern for over 30 years, the current situation is dramatically different from that of previous decades.

Table 8.1   Structure of world employment, 1965 and 1989/91

| | Percentage of total employment in | | | | | |
| | Agriculture | | Industry | | Services | |
Region	1965	1989/91	1965	1989/91	1965	1989/91
World	57	48	19	17	24	35
Industrialised countries	22	7	37	26	41	67
Developing countries	72	61	11	14	17	25
East and Southeast Asia	73	50	9	18	18	32
Sub-Saharan Africa	79	67	8	9	13	24

Source: International Labour Office, 1995

In the prosperous years of the 1960s, Third World unemployment was attributed to the lack of modern industrial jobs for the rapidly growing urban labour force. Particular concern was directed towards large numbers of rural migrants searching for new work opportunities in the city (Edwards, 1974). Solutions to unemployment were seen to lie in increasing industrial output through policies of modernisation. In the 1980s, however, employment creation was severely restricted by the debt crisis and austerity measures implemented by the International Monetary Fund and the World Bank. As Third World manufacturing decreased under global recession, the future applicability of industrial-based employment programmes has been questioned. Following a marked deterioration of employment prospects and increased poverty, the current challenge is to create sustainable employment for future generations in the form of secure jobs free from exploitation and hazard.

A central problem in the developing city is the lack of well-paid employment opportunities for the majority. In an urban context, most jobs are concentrated in the manufacturing and service sectors, with the latter increasing in importance. Table 8.1 shows the growing importance of service employment throughout the developing world since 1965. Another major formal sector employer in large cities is the public sector, and although declining, it still accounts for 40 per cent in many African cities (Turnham, 1993). Recruitment is frequently through established social and political networks, rather than official channels. High status, modern occupations are frequently allocated by a system of patronage, although the recent privatisation of public sector services across the developing world has led to a restriction on the intake of graduates, a group who are now less able to migrate to the west in search of employment.

Employment opportunities are often heavily skewed to primary and secondary cities due to the concentration of wealth, technology and modern enterprise. Primary cities frequently account for 50 per cent of a country's manufacturing and service sector jobs. Mexico City, for example, represents the heaviest industrial concentration in the country and has the highest population growth.

However, many urban workers are still unable to secure a job in the modern or formal economy. Since the 1980s, increasing international competition, mechanisation, and external pressures from structural adjustment, have resulted in the widespread closure of businesses in all sectors of the economy. In an experience which bears some similarities to western economies, it is the loss of unskilled and semi-skilled manufacturing jobs and the creation of low-paid, part-time service sector employment which has altered the structure of the labour market. In many cities, male and youth unemployment has been growing whilst many female workers have found jobs in newly created service employment.

As a result, urban labour markets have become increasingly fragmented under a hierarchical structure which constitutes an elite sector of senior professionals, and a middle group where employment is dominated by public sector executives, managers and entrepreneurs in domestic and foreign industry, services and finance. The lower spectrum of this middle group sometimes includes public sector professionals, although reductions in public wages have increased the incidence of poverty amongst teachers and nurses, for example. Beneath this middle group remains the largest concentration of the urban labour force which constitutes semi-skilled and unskilled manual labour and low-level administrative and service occupations, in both the formal and the informal sectors.

Urban labour markets: growth and inequality

Urban labour markets in developing countries are highly fragmented and, although individual structures differ, there are a number of general characteristics which make them distinct from cities in the developed world. Firstly, as the number of people searching for work depends on the size, gender and age structure of the population, urban labour forces in developing countries have grown at unprecedented rates of around 2 per cent per annum, due to population growth and rural–urban migration. Excluding China, the population of the developing world is increasing by 2.3 per cent per annum compared to 0.3 per cent in developed nations, resulting in a youthful population (UNDP, 1996; Todaro, 1995).

There are increasing disparities in labour force growth rates between countries, with deceleration occurring in the Asian NICs and Latin America's southern cone, but increased growth in sub-Saharan Africa, Bangladesh and Pakistan. Labour force participation rates, which range from 40 to 80 per cent of a country's working population, are strongly influenced by female participation rates. Between 1955 and 1985, labour force growth was partly attributed to the increased participation of women in the formal labour force, due to the 'feminisation' of service employment and the pressures of structural adjustment on household incomes (Chant, 1992; Gilbert, 1994). Ultimately, figures vary according to the 'visibility' of female labour between countries.

Secondly, rural–urban migration has overstretched the coping mechanisms of most cities, a phenomenon which has contributed to the development of a large informal sector in many cities. Historically, regional or national production centres attracted rural workers in search of jobs in the modern sector, a trend which was seen to replicate western development patterns. Following the Industrial Revolution, new employment opportunities were created which provided jobs for redundant rural labour. It was argued by Hirschman (1958) that this process would be repeated in cities across the developing world. In the 1950s and 1960s, supporters of the modernisation approach, such as Arthur Lewis (1955), viewed cities as growth centres which could absorb unlimited supplies of labour from rural areas. In reality, rapid urbanisation and population growth over 25 years, combined with massive rural–urban migration and decreasing agricultural productivity, led to a growth in urban unemployment and underemployment as labour markets failed to provide modern sector employment as envisaged by advocates of modernisation.

Thirdly, there is a close relationship between levels of unemployment, poverty and inequalities in the distribution of income. A Eurocentric perspective on employment prospects in the developing world often portrays a negative image of Third World workers, identifying them as low-paid, manual labourers operating in hazardous conditions. Whilst such occupations do indeed exist in large numbers, many cities have a dynamic modern sector which offers well-paid occupations and opportunities. It is the inaccessibility of modern-sector careers to a large proportion of the workforce which creates the mismatch between opportunities and aspirations.

In addition to their lack of access to education, recognised qualifications and formal training, many workers are also disadvantaged by a rigid division of labour shaped by gender, race, ethnicity and class. In Malaysia, for example, the state encourages the development of Muslim businesses through active discrimination, particularly in national contracts such as transport. In Kuala Lumpur, Malay workers frequently find enhanced employment opportunities relative to members of other groups. In the Caribbean, where colonial practices once dictated social structure, the forces of gender and ethnicity interplay to produce a hierarchical division of labour which still values 'whiteness'. In India, the caste system still dictates that 'untouchables' are restricted to occupations such as leather beating or pot making. Until recently, the apartheid system in South Africa relegated coloured workers to low-status occupations, regardless of their education and level of skills.

In relation to gender, many studies have highlighted the channelling of female workers into unskilled and low-paid 'women's jobs' as a result of social stereotyping (Scott, 1994; Elson, 1995). Feminist perspectives have identified women's reproductive role, particularly in relation to domestic work, as responsible for their subordinate position in the labour market (Beneria and Sen, 1981; Rogers, 1980). Colonialism is seen as being responsible for undermining the value of women's work in many developing societies, as it introduced the concepts of the male breadwinner and the female housewife (Mies *et al.*, 1988). As documented by Redclift and Sinclair (1991), orthodox economic justification for women's low pay and job status stems from the fact that they are all too often perceived to have lower levels of skills, aptitudes and/or education ('human capital'). Women are also regarded as constituting a more passive workforce, which is less likely to form resistance to poor work practices, although this is a view which is gradually being challenged. Gender stereotyping restricts women's employment in both the informal and formal sectors to poorly paid, low status jobs such as domestic work, clothing production, catering and, more recently, assembly work in Export Processing Zones, where women are employed for their dexterity and ability to undertake repetitive tasks (Chant and McIlwaine, 1995b; Safa, 1990; Pearson, 1994). Proponents of modernisation theory argued that industrialisation would advance women's position in the labour market through new work opportunities. It is interesting to note that when occupations become increasingly feminised, as is the case with EPZ manufacturing, they tend to become de-skilled.

A high percentage of the female workforce is also employed in the informal sector (Scott, 1994, 1995; Massiah, 1993) as a result of their exclusion from formal employment. Unequal access to education, training and credit is often compounded by women's reproductive responsibilities which restrict work outside the household. As in the formal economy, women are commonly found in low-waged informal occupations such as food preparation, petty commodity production, street trading or working in subcontraction outlets (Moser, 1978) (see Plates 8.1 and 8.2, overleaf). In Latin America, 25 per cent of the female workforce are domestic maids. In Brazil, domestic service accounts for 56 per cent of the jobs black women undertake, as opposed to 24 per cent of the jobs occupied by other groups (Cubitt, 1995). The failure to value women's work across all sectors of society has major implications for development, particularly in the African and the Caribbean contexts where female-headed households are prominent (Ellis, 1986; Momsen, 1993).

Fourthly, urban labour markets are shaped by their role in the wider national and international economy, where reliance on international capital can lead to instability. Changes in world production since the 1960s have influenced employment patterns. Roberts (1991) argues that Guadalajara, which is now Mexico's third-largest centre for industrial production, grew as a result of attracting both small and medium-sized domestic firms from the nineteenth century. In contrast, Mexico City's development was heavily influenced by foreign investment, which today has resulted in the proliferation of capital-intensive branches of multinationals and small subcontraction firms. In relation to employment, Guadalajara's structure proved more successful during the 1980s recession, particularly in its ability to increase formal manufacturing employment in smaller outlets. During the 1980s, Mexico City's industrial labour market was severely hit by global recession, leading to widespread redundancy amongst factory workers.

According to Simai (1995), the main global processes shaping employment in the 1990s include technological change and 'jobless growth', the growth of both youthful and ageing job seekers, the increasing feminisation of the labour market, and the environmental impact of job creation as the result of global concern for sustainable development. Furthermore, the possibility of full employment is constrained by internal and international imbalances, diverse social interests and power groups, limited wealth and resources and international

Plate 8.1   Informal sector fruit and vegetable sellers in Kingstown, St Vincent (photo: Rob Potter).

Plate 8.2   Informal sector vendors in Bolivia (photo: Astrid Bishop).

competition in a post cold-war era which has seen an increase in jobless economic growth.

Urban unemployment and underemployment

Unemployment in the Third World is frequently viewed as a symptom of unequal and inadequate development, but attempting to identify the 'unemployed' remains a highly controversial task. Firstly, difficulties arise as attempts to measure categories of employment are generally constructed around western concepts of work (Pahl, 1988). Western theories of urban economic structure and labour force segregation have failed to adequately analyse and explain the structure of the urban labour force in developing countries. One issue which is widely agreed upon is the pressing need to work. A subject which is less clear centres on the definition of the boundary between work and employment in the developing world context.

Work undertaken outside the realms of officially accounted employment is prevalent in the urban economies of developing countries. Much productive work is undertaken outside the confines of the formal waged economy, indeed many household productive and reproductive activities are unremunerated. Informal activity, self-employment, casual work and home-based production have provided alternatives to formal employment for decades. It is important, therefore, to expand western notions of work to include a wider range of social and economic activities which, due to their unregulated nature, often have most salience in urban areas.

Secondly, there is the problem of definition (Gilbert and Gugler, 1992). The ILO defines the 'unemployed' as those who are without work but who are actively seeking it (Hussman, 1990). It is widely argued that the true unemployed are not amongst the urban poor. As far back as 1968, Myrdal argued that unemployment is a luxury few can afford in such circumstances (Myrdal, 1968). The 'luxury unemployment thesis' (Udall and Sinclair, 1982) argues that due to the absence of social security, only educated professionals, such as white-collar workers, public sector executives, or possibly new entrants to the labour market who are still able to rely on family support, contribute to the true unemployed in most developing countries. As a result, Udall and Sinclair (1982) argue that the unemployment figures published by official agencies are meaningless and inappropriate, for they fail to provide a clear picture of the true structure of employment. Turnham (1993), however, argues that real unemployment is a global problem for the 15–24 age group, both educated and uneducated, and is increasingly becoming a political dilemma.

As a result, low participation rates in the formal economy, along with measured unemployment at under 10 per cent for many countries, means that a substantial proportion of the working population is not accounted for. Table 8.2 (overleaf) clearly highlights this situation in relation to Latin America. Such a situation is often due to the unsuitability of the criteria used for inclusion and exclusion of individuals from the official labour force, which largely exclude informal employment and household work. As Turnham (1993)

Table 8.2   Urban unemployment in selected Latin American countries

Country	Percentage unemployment				
	1980	1985	1989	1990	1991
Argentina	2.3	6.1	7.6	7.4	6.5
Bolivia	7.1	5.8	10.2	9.5	8.1
Brazil	6.3	5.3	3.3	4.3	5.0
Chile	11.8	17.0	7.2	6.5	7.9
Colombia	9.7	14.1	9.9	10.3	10.3
Costa Rica	6.0	6.7	3.7	5.4	5.0
Ecuador	5.7	10.4	7.9	–	–
Mexico	4.5	4.4	2.9	2.9	2.6
Peru	7.1	10.1	7.9	8.3	–
Venezuela	6.6	14.3	9.7	10.5	10.9

Source: Cubitt, 1995

argues, unrecorded production absorbs as much labour time as all combined activities in the recorded sectors of the economy. Goldschmidt-Clermont (1990) estimates that household work accounts for 30–50 per cent of GDP in many developing nations. Household activities, which are often undertaken by women and children, include household enterprise, house construction and maintenance, housework and child care, and urban food production (Drakakis-Smith, 1990). The failure to include these activities as productive employment often results in unrealistically low female participation rates, and provides a distorted view of the relative importance of formal employment in many developing countries.

Due to the inadequacy of the term 'unemployment', it is now widely recognised by researchers and academics that *underemployment* provides a better indication of the employment problem. Gilbert and Gugler (1992) define underemployment as the underutilisation of labour in three distinct forms. The first and most common form occurs where workers are so numerous that many are less than fully employed, or work fewer hours than they would like. Street trading provides a good example. Secondly, underemployment can be related to fluctuations in economic activities during the day, week or month when there is little or no work. Thirdly, there is 'hidden underemployment' where people are employed even when there is insufficient work to keep them fully occupied. According to the ILO, 'visible underemployment' refers to those who have worked less than the normal duration, 'disguised underemployment' to workers whose skills have been underutilised, whilst 'potential underemployment' accounts for those whose productivity is inadequate (Hussman, 1990; ILO, 1995). Underemployment has also been applied to workers who earn less than the average wage, as defined by Sabot (1979) in the context of Tanzania. There are also what are referred to as 'discouraged workers', those who are not working but who would like to, but have given up due to the lack of opportunity (ILO, 1995).

A large proportion of the underemployed are seen to constitute the informal sector which provides part-time, casual and unproductive jobs for the urban poor. As the next section will show, however, the informal sector is not just a haven for marginalised workers, but a positive provider of productive employment opportunities, albeit one that is associated with certain difficulties and problems.

Alternative forms of work: the urban informal sector and the experiences of the working poor

Early perceptions on informality

It is in the light of underemployment and the lack of modern sector jobs that theoretical debates have focused on the structure of the Third World urban economy. Many theories of labour force segregation, with the exception of recent feminist contributions (Stitcher and Parpart, 1990; Redclift and Sinclair, 1991), are not applicable to the developing world in their entirety. Interest in the existence of a 'traditional', or as it was often referred to, 'marginal', employment sector, gained momentum in the 1950s and 1960s, with the rise of modernisation theory. As detailed in Chapter 3, the modernisation school argued that developing countries should follow the path of developed nations, by pursuing the goal of urban-based industrialisation (Lewis, 1954; Rostow, 1960; Hirschman, 1958). It was argued that the traditional or backward economic sector, which incorporated a large pool of casual labour, would thereby be eradicated once prosperity increased. Structural problems, such as unemployment and underemployment, were seen as temporary impediments which would be overcome by industrial development (Hirschman, 1958).

Dualism referred to the co-existence of an 'advanced', or 'modern', sector with a 'backward' or 'traditional' one within the Third World urban economy. Supporters of this approach regarded the traditional economic sector as the major barrier to development. This set the background to one of the most influential theories of development in the 1950s, provided by Lewis's ideas on unlimited supplies of labour, as mentioned in the previous section (Lewis, 1955). Lewis combined his knowledge of the West Indian experience with classical economic theory, to outline an economic situation where unemployment was high, and the ability to absorb the increased labour force had been exhausted. Lewis's solution to the problem was industrialisation by invitation, a model which was pursued throughout the developing world during the 1960s, as outlined in Chapter 3.

The heyday of the dualistic approach was the 1970s, when growing criticisms of marginality (Reynolds, 1969; Quijano, 1974) focused attention upon the problems of employment and economic structure. The concept of marginality implies an existence on the periphery of the economy or society, and has been largely discredited, as it implies that many people live 'outside the system' (Perlman, 1976; Lomnitz, 1977). Since then the concept of structural marginality has been used to assess the degree of participation within the system and it is increasingly appreciated that people are disadvantaged by the system, rather than being external to it.

The modern sector was identified as a dynamic contributor to the economy, and as a provider of well-organised labour opportunities, characterised by large-scale industry, the public sector and other regulated businesses. By contrast, the backward sector referred to those activities which were unregulated by government, and which were seen as inefficient, small-scale and unproductive. These sectors were identified as separate entities, operating independently of one another. Although dualism has been widely criticised for its failure to recognise the linkages which exist between sectors (Breman, 1976), its merit lies in drawing greater attention to the existence of an informal employment sector. Santos (1979), developed this model further by stating that the sectors were inextricably linked within the economy. Early perspectives on the development of the informal sector closely followed the path of development theory, with many discussions of the informal sector taking the dualist school of thought as their point of departure.

Perspectives on the nature of the informal sector

> They have simply renounced legality. They go out on the streets to sell whatever they can, they set up their shops, and they build their houses on the hillsides, or on vacant lots. Where there are no jobs, they invent jobs, learning in the process all they were never taught. (de Soto, 1989: xix)

The informal sector refers to unaccountable and unregistered activities which are found in most countries of the world. Despite difficulties in measuring the exact scale of the informal sector, it has been estimated that 50 per cent of the labour force in developing countries work within it, as opposed to 3 per cent in developed nations (Thomas, 1992). According to de Soto (1989), 48 per cent of Peru's active population work in the informal sector, contributing nearly 40 per cent to GDP. Furthermore, he estimates that of 331 markets in Lima, 274 have been built by informal black marketers, who also control 95 per cent of the transport system.

The informal sector is heterogeneous, with respect to both its activity and its workforce. The majority of goods and services available in the formal economy have similar informal versions which serve both low-income communities and the international economy. Within the informal sector there are a range of small-scale producers, retailers and service providers, many of whom operate beyond the reach of government. Work ranges from rag-picking in Bombay or rickshaw driving in Hong Kong, to the operation of informal garages in Port of Spain. The degree of informality differs greatly according to a number of decisive characteristics. These include the type of commodities produced, the method of production, the type of ownership, market strategy, the number of additional people employed and geographical location.

As shown in Table 8.3, popular primary and secondary sector production includes market gardening, building, artisanal craft production, manufacturing

Table 8.3 A functional typology of typical informal sector operations in developing countries

A. Small-scale production

(i) *Primary activities*
Market gardening
Urban farming
Construction

(ii) *Petty commodity production* (secondary)
Food processing
Home-production of hot food
Garments
Crafts
Jewellery and trinkets
Shoes and leather products
Household goods
Electrical and mechanical items
Specialised production (e.g. festivals, alcohol)

(iii) *Tertiary activities*
Printing and artwork
Office equipment
Computing and software

B. Distributive trades and tertiary enterprises

(i) *Distributive trades*
Processed food trading (nuts, snacks)
Unprocessed produce (fruit and vegetables)
Commercial food trading (Chiclet, Coca Cola)
Suitcase trading (imported items)
Hot food and drinks
Clothes, shoes and leather goods
Jewellery and cosmetics
Newspapers
Household items
Music and electrical items

(ii) *Tertiary services*

Daily service providers

Laundry
Domestic
Shoe cleaning and repair
Hardware repair
Motor vehicle servicing
Taxi-driving and transport
Maintenance and gardening
Odd jobs, e.g. car cleaning
Bottle and waste collecting

Specialised services

Tourist guides
Car park attendants
Car, home rentals
Residential lodgings
Secretarial, clerical
Legal and medical
Beauty services/hairdressing
Distribution/storage
Begging
Protection

Source: Lloyd-Evans, 1994

production, and food and alcohol-related industry. Tertiary enterprises, and distributive trades, include transport provision, petty trading, street hawking, catering, domestic and restaurant services, agents and dealers (see Plate 8.3, overleaf). Other informal service providers include musicians, launderers,

Plate 8.3   Selling belts and ties from a mobile trolley in Hong Kong (photo: Rob Potter).

barbers, photographers, vehicle repairers and other maintenance workers, brokers and middlemen.

There is a large body of literature which attempts to define and explain the informal sector (ILO, 1972; Bromley and Gerry, 1979; Tokman, 1989; de Soto, 1989; Thomas, 1995). Despite this contribution, investigations into the nature of informal sector activity have often been obscured by theoretical debate over its definition. As Portes *et al.* (1991: 1) comments, 'the informal sector has come to constitute a major structural feature of society . . . and yet the ideological and political debate surrounding its development has obscured comprehension of its character'. Although the notion of a clear-cut formal–informal sector dichotomy has largely been rejected in favour of recognising that there are complex interactions between different types of economic activity, the distinction is still used to identify the different ways in which people earn a living. According to Castells and Portes (1991), the informal economy is a commonsense notion, the moving boundaries of which cannot be frozen by a strict definition.

Despite its widespread recognition as a 'new concept', following Keith Hart's work in the 1970s, the existence of informal sector activity had been documented since the 1950s (Hart, 1973; ILO, 1972). Prior to the 1960s, there had been a number of studies dealing with irregular economic activities (Lewis, 1955; Simpson, 1954). Beggars, hawkers and small-scale operators were commonplace in the fast-growing developing cities of that time. In a study of street begging in Kingston, Jamaica, Simpson (1954) identified street workers as

Table 8.4   The enterprise-based approach to the informal sector

Characteristics of the informal sector	Characteristics of the formal sector
Ease of entry	Difficult entry
Reliance on indigenous resources	Reliance on overseas resources
Family ownership of enterprise	Corporate ownership
Small-scale operation	Large-scale operation
Labour-intensive method of production	Capital-intensive production
Use of traditional technology	Use of imported technology
Skills acquired outside formal system	Formally acquired skills and qualifications
Unregulated and competitive markets	Protected markets (tariffs and licences)

Source: International Labour Office, 1972

social failures. Following Oscar Lewis's (1966) article 'The culture of poverty', negative views were projected of informal operators.

Lloyd (1979) referred to the sector as that of the 'lumpenproletariat', consisting of workers who had rejected the dominant values of society and who had turned to crime, prostitution and begging. Also referred to as the 'marginal economy', the sector was seen to include a vast range of economic activities which were poorly organised, difficult to enumerate and invisible in the official census (Breman, 1976). Another major influence in the study of informality has been the International Labour Office. In 1964, the ILO adopted the Employment Policy Convention (No. 122) which encouraged governments to follow a goal of active full employment. The ILO's World Employment Programme, at the 1967 International Labour Conference, suggested that a number of country-based pilot studies should be carried out. The first country mission was Colombia in 1970, which was followed by Sri Lanka in 1971 and Kenya in 1972. These studies questioned the applicability of the concept of unemployment, as defined in developed nations, to developing countries, where there is no social security system. The mission to Kenya changed the focus from unemployment to the employment of the 'working poor' (ILO, 1972: 9), and the sector was dubbed 'informal' by Hart (1973).

There have been numerous attempts to establish criteria for the recognition of informal activities. A number of early studies, commonly referred to as the 'enterprise approach', attempted to define the informal sector through a list of characteristics which distinguished between informal and formal activity (ILO, 1972; Weeks, 1975; Davies, 1979). Davies (1979) argued that informal sector activities are characterised by their ease of entry, small-scale, intensive use of labour, and family ownership, as shown in Table 8.4. By contrast, formal sector activities required trained workers for their large-scale, capital intensive, corporate businesses. Santos (1979), Breman (1976) and McGee (1979), amongst others, criticised this approach for its failure to recognise the numerous inconsistencies which exist. For example, many informal businesses have their own entry requirements. More recent work by Tokman (1989) and

Rogerson and Hart (1989) provides further evidence to suggest that the structure of informal enterprises does not always differ from that of its formal counterparts.

Other researchers have examined the idea that the fundamental characteristic of informal employment is that it is based on self-employment (Singer, 1970; Hart, 1973; Scott, 1979, 1986b). Scott (1979) advocates that, once again, no clear-cut dichotomy exists, as there are a number of intermediate forms of operation which exist between waged work and true self-employment. Increasingly, waged labour is becoming a common activity in the informal sector, particularly in the form of contractual arrangements with corporate firms.

Following the self-employment approach, a popular perspective focused upon the mode of production as the decisive factor in identifying informality (Davies, 1979; Moser, 1978, 1984). It was argued that the formal sector differs from the informal sector, because it represents a different mode of production, the informal sector being non-capital intensive and owned by those who operate such activities. In the formal sector, the means of production are capital intensive, and they are privately owned by a small elite class. This approach led to the widespread notion of the petty-commodity producer.

Moser (1978) argued that the informal sector is a subordinate mode of production, exploited by capitalism. Based on both the Marxist concept of 'petty commodity production', and on the theory of unequal exchange, the radical school reinterpreted the informal sector as a reserve army of labour (Castells, 1977; Connolly, 1985). Economic surplus generated in the informal sector is transferred to the formal sector through two major mechanisms. Firstly, the informal sector lacks access to the basic resources of production, which are monopolised by the formal sector, Secondly, the informal sector is forced into a position whereby it must pay higher prices for its purchases, but can only ask low prices for its outputs. Therefore, the informal sector is limited by its relationship to the formal sector, and the sector as a whole is said to be 'involuting'.

The informal sector is seen to be exploited by the capitalist economy in a 'help the poor without threatening the rich' strategy (Bromley, 1978: 1036), which eradicates the need for the state to provide jobs for its labour force. Castells (1977) observed that the informal sector is similar to informal housing areas, which provide a cheap, easy solution to the housing crisis, a point covered in the previous chapter. The informal sector is seen as a contributor to the accumulation of the wealth of core states via the employment of informal workers by multinational firms. A prime example of the interactive nature of the formal and informal distribution networks is illustrated by the role undertaken by street traders. Many formal sector goods such as Coca Cola are sold via an informal distribution system, where street traders buy direct from the distributor to retail on the streets. Safa (1981, 1986) and Scott (1986a) have documented the fact that these unregulated activities play a critical role in subsidising the costs of multinational operations in the Third World, particularly in their use of cheap female labour.

As previously documented, the search for a more widely accepted approach has focused upon the identification of linkages between the informal and formal sectors (Santos, 1979; Sethuraman, 1976, 1978; Tokman, 1989). Primarily, the informal sector contributes to the growth of the economy. By providing cheaper goods and services to poor communities, it also transfers capital to the formal sector, which is seen as an undesirable process which furthers uneven development. In addition, the informal sector takes on a vital social security role at a time when public spending is decreasing. Following this line of thought, the concept of an informal–formal continuum, along which all economic activities take their place, has gained much support. Sethuraman (1981) placed many small-scale informal enterprises, such as hawking and petty commodity production, at one end of the continuum, and suggested that formal enterprises, such as TNCs, would be placed at the other end. He also suggested that as an informal activity grew, it would occupy a different position along the continuum. As a result, the informal sector is seen to perform a number of differing roles in the urban economy.

Marginal workers or entrepreneurs? Experiences of informality

The informal sector is ambiguous, in that it encompasses wealth and poverty, productivity and inefficiency, exploitation and liberation. As discussed previously, debates still focus on whether the sector is a haven for structurally marginalised workers or productive entrepreneurs (Portes and Walton, 1981; Portes et al., 1991; Castells, 1982). As informality is becoming increasingly widespread, its participants emanate from a wide range of social groups, and they enter the informal sector for different reasons. Following the existing literature (Portes et al., 1991; Thomas, 1992, 1995), it is possible to identify four main sectors of informal activity which attract different groups of workers, as highlighted in Figure 8.1 (overleaf) derived from Lloyd-Evans (1994).

Firstly, there is the **subsistence sector**, where goods and services are produced by the household primarily for its own consumption. The well-known activities of housework, the use of unpaid family labour and reproduction fall into this category (Moser, 1984; Chant, 1987). Small surpluses may well be sold in the local market, but the prime motive of production is subsistence, rather than profit.

The second category consists of **small-scale producers and retailers**, who primarily produce or sell for the acquisition of a wage. They are often self-employed, and have established their own business in order to provide themselves with work. Typical workers in this category are street traders, artisans, food sellers and some home-based producers. In the Caribbean, the familiar activities of the female 'higgler', who sells vegetables and petty commodities on the streets, would be included here (Lloyd-Evans and Potter, 1993; Le Franc, 1989). Some small family enterprises would also fall into this category.

In addition, many households receive both informal and formal incomes. Workers, often female, utilise informal employment to subsidise low formal wages, either on a part-time basis or by another member of the household.

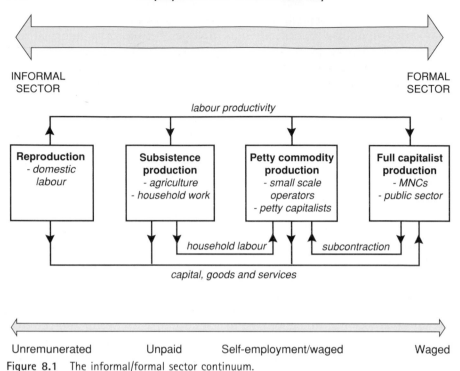

Figure 8.1　The informal/formal sector continuum.

Others are 'recessional workers', who participate in informal work in times of hardship, but discard their jobs when household income levels rise. Many street traders, clothing producers and daily service providers engage in the sector for this purpose. According to Sethuraman (1981) and Tokman (1989), an operator's motive for entering the sector is an important consideration for determining the degree of informality of a business. The creativity and entrepreneurial spirit of many informal workers have been channelled into the development of small, productive, retail and service enterprises which make use of the avoidance of tax and other legislation. It is important to note that self-enhancement and an improved standard of living are the main motives of workers in this group, who hope the growth of a small business will assist in their social mobility.

A third category is one which is increasingly forgotten, but which accounts for a large proportion of the informal sector, particularly in Latin America. The main occupants of this sector could be called the '**petty capitalists**', as their main aim for establishing an informal business is the extraction of profits, which may be greater than in the formal sector (de Soto, 1989). Social climbing is not the only motive behind this category of worker, who use informal means of production for the purpose of increasing profits. The avoidance of taxes and employment legislation and regulations such as minimum wages, hours and conditions, attract many successful entrepreneurs to the informal sector. Petty capitalists will frequently employ waged labour, often at low rates in sweatshops in order to produce goods at low cost. These businesses are evidence of the fact that large profits can be made within an informal unit, and support the argument that the informal sector is not entirely made up of the 'casual poor' (Bromley and Gerry, 1979; Teltscher, 1994). In India, employment in the informal manufacturing sector accounted for 75 per cent of total manufacturing employment in 1990.

A large proportion of petty capitalist businesses will be contracted to large formal sector firms, and it is within this type of business that the greatest exploitation is found. Modes of production range from traditional labour-intensive processes to capital-intensive production using imported technology. Workers, many of whom are women, have no protection and are exploited by the informal system, a point that is further developed in Chapter 9. Many women are forced to take low-paid informal sector occupations as their lack of education and skill, combined with reproductive responsibilities, restricts them from the formal economy (Redclift and Sinclair, 1991; Momsen and Kinnard, 1992; Scott, 1994). In a similar manner, many large enterprises will employ out-workers to undertake menial tasks in their homes. The clothing and electronic industries are prime examples of this type of 'informal' employment, and have been well documented by several authors (Safa, 1990; Lawson, 1992; Mitter, 1994).

A fourth category of informality refers to those activities which are **criminal and socially undesirable**. Criminal activities such as drug dealing and smuggling, theft, extortion and protection are all part of the informal sector, and are becoming increasingly important in global capital accumulation. Prostitution is also illegal in many countries of the world, and is most prevalent in Asia (Lee, 1991; Hall, 1992). The international drug trade, for example, is a major employer of small-scale producers, dealers and smugglers in countries like Bolivia (Blanes Jimenez, 1991; Thomas, 1992).

The recognition of a wide range of activities reflects the recent changes in attitudes towards the informal sector. No longer seen as a homogeneous group of marginalised activities, the informal sector is now being encouraged and supported. Research has identified the benefits of informal employment, which include local control over the forces of production through self-employment, higher incomes, independence, and tax avoidance. The informal sector represents an opportuntity for income generation, which is characterised by a lack of regulation by government institutions. If supported, it is argued that many informal businesses have the capacity to grow and expand. De Soto (1989) argues that the informal sector is growing in popularity amongst many social groups as it represents a move away from inefficient state bureaucracy. In his view, the informal sector can be seen as a generic entrepreneurial environment for Latin America, which brings together bonds of popular solidarity, creativity and flexibility.

Despite the positive benefits of informality, there is still concern over the inability of certain social groups to gain access to this entrepreneurial environment, as economic and social cirumstances still confine certain groups to low-paid occupations, based on the unequal distribution of resources in both the local and international economy. As this chapter stresses, the unregulated informal sector is a major employer of the world's most vulnerable workers, often women and children, who serve as a cheap source of exploited labour power. Informal workplaces rarely conform with labour legislation laws or environmental regulations, circumstances which will receive detailed examination in Chapter 9.

Recent thought has recognised the need to support informality, without attempting to prescribe its boundaries (Tokman, 1989; Castells and Portes, 1991). Encouragement of the informal sector, where people's purposeful and intelligent energies are used for productive purposes, is now widely seen as an essential component of successful development policies. National governments are attempting to 'formalise the informal economy' through projects which offer credit and business training to workers in micro-enterprises, whilst Non-Governmental Organisations (NGOs) are implementing income-generation schemes to assist the most disadvantaged (Massiah, 1989; Simon and Birch, 1992). Access to credit and resources in the form of credit associations and community organisations is a particularly important step for female informal producers who lack formal assistance. For example, women constitute 18 per cent of the self-employed in the developing world, but in Latin America they receive only 11 per cent of formal credit programmes (UNDP, 1996). Supporters of the informal sector are concerned that given current global pressures, informal operators are more likely to become exploited disguised waged workers, rather than successful entrepreneurs, unless greater assistance is given to harness their potential.

Work, globalisation and the city: the search for cheap labour ___

Urban workers and the New International Division of Labour

Work and employment patterns are undergoing a multi-dimensional restructuring which represents part of the process of globalisation which, as noted in earlier chapters, is contributing to uneven development and social polarisation rather than uniformity (Allen and Hamnett, 1995; Massey and Jess, 1995; Johnston *et al.*, 1996). Economic globalisation is evolving as a result of the internationalising of trade, production and consumption through the development of global factories and chains of flexible corporations. Just as capitalism rewards owners of capital relative to labour, so it empowers multinational corporations over national governments in relation to market domination, labour legislation and employment policy. As discussed in Chapter 3, there is a global market for labour which has forged stronger interdependent links between developed and developing nations (Thrift, 1986; Dicken, 1992; Chandra, 1992; Sklair, 1994b; Daniels and Lever, 1996). The promotion of export-oriented manufacturing in developing countries, combined with the demand for cheaper labour by multinational companies and increased technology, provided the impetus for what has become known as the New International Division of Labour (NIDL).

Since the 1980s, problems of structural employment and domestic recession have resulted in increasing competition between Third World governments in order to attract foreign direct investment (Frobel *et al.*, 1980). Flexible specialisation and 'just-in-time' manufacturing techniques have led western companies to search for cheap, unregulated labour. As production processes became increasingly fragmented and homogenised, many sub-processes could be carried out by semi-skilled or unskilled labour. In 1994, multinational firms employed 12 million workers in the developing world in a range of export industries. The breakdown of traditional economic and social structures in many countries as a result of poverty, debt and dependency has led to a vast supply of cheap labour available for work.

Reflecting neo-liberal strategies of economic development, the global economy is becoming more 'informal' in terms of the conditions and practices of employment (Castells and Portes, 1991; Sassen, 1988). In times of increasing global competition, part-time, contractual and low-paid work is becoming the norm, particularly for migrant workers (Cohen, 1987). Global deregulation and expansion of the private sector without restriction is leading to the informalisation of employment throughout the world. Flexible specialisation and 'just-in-time' are synonymous with the search for cheap, flexible and non-unionised labour forces, which are characteristic in the poorer economies of the world. This process has also led to an increasing feminisation of the labour force. The 'Benetton Model', as it has often been called, refers to the operation of firms which subcontract small workshops in developing countries to produce their clothing. Benetton, like many other clothing producers, do not have production plants in the UK or Western Europe but instead use a flexible chain of producers around the globe, often employing low-paid workers in sweatshop conditions.

The growth of regional trading arrangements, previously mentioned, has resulted in a change in North to South trade. Harcourt (1994a) argues that whilst 'free trade' might allow for accelerated economic growth and interdependence in advanced industrial economies, this happens at the expense of countries which have experienced uneven development. As will be discussed in Chapter 9, the negative effects of deregulation such as environmental pollution, exploitation and the social exclusion of certain groups and communities, are largely felt in the countries of the South. As a result, employment in light industry (clothing and textiles, for example) is predicted to increase in countries like India, but at what cost? It is argued that children and female workers are likely to occupy these newly created positions. Furthermore, the IMF and World Bank promote foreign direct investment as a solution to underdevelopment, a process which has major implications for developing cities and their workforces.

Export industries, subcontraction and the household: towards the feminisation of the workforce

A major part of the NIDL has been the development of EPZs in countries such as India, the Philippines, Mexico, Panama, Puerto Rico, Jamaica and

Barbados, specialising in light manufacturing and data processing. Specific requirements for a 'cheap and passive workforce' have led to the feminisation of many jobs in EPZs and Free Trade Zones (Pearson, 1994; Standing, 1989; Chant and McIlwaine, 1995b), where foreign firms are offered incentives to invest. An estimated 75 per cent of EPZ workers globally are female, with employers holding a preference for unmarried women under 25 years. Jobs in clothing, textiles, electronics and data processing are most common. In Asia, half a million workers in EPZs are concentrated in Singapore and South Korea alone. Women work long hours for poor wages under harsh conditions (Nash and Fernandez-Kelly, 1983). Most enclave firms insist on non-unionisation and the deregulation of extant national labour laws. Safa (1990) states that, although this work does little to raise the status of female employees, jobs in data processing are seen to be more desirable than traditional alternatives such as domestic service. As a result, data processing for companies like American Airlines is a popular occupation amongst women in Barbados, despite the low wages and conditions (Clayton and Potter, 1996).

Another recent phase in the NIDL links the urban household and informal sector to the international system through a system of subcontraction. With the exception of EPZs, multinationals are increasingly abandoning Fordist-type factory production in favour of profit maximising subcontraction, which minimises labour costs. Many global clothing companies are employing female workers at home to produce consumer goods. The emergence of this new casual workforce resembles nineteenth-century European cottage industry, where workers are unprotected by legislation, paid on a piece rate, and have to meet stringent quotas. Here, masses of informal sector workers, often women, are employed in small informal workshops producing goods for export to the West.

The increasing number of poor, informal, waged labourers in developing countries represents one of the regressive aspects of the development of informal sector employment. It has been argued by Safa (1990) that the existence of a large informal sector attracts foreign companies to take advantage of cheap labour in the Caribbean. In this way, workers are specialising in a single part of the production process that has connections throughout the global economy and contributes further to western capital accumulation.

Child labour: a necessary evil?

One of the most serious consequences of neo-liberalism is the increase in child labour which has occurred across the globe, a trend which results from increases in both supply and demand. Poverty and deprivation increase the need for children to seek paid labour, a trend which is often being met by a rapidly rising demand for cheap child workers within the global economy.

The world's child workforce is estimated at 200 million, with a predominant concentration in developing nations (Action Aid, 1992). Asia has the largest percentage of child workers, with India accounting for 44 million alone, compared to Africa's total of 20 million (Fyfe, 1994). In Latin America it

Plate 8.4   Children turning over rubbish at the rear of a major hotel in Delhi, India
(photo: Rob Potter).

has recently been estimated that between 10 and 25 per cent of all children
are working (UNDP, 1996). Child labour takes on many forms, from paid
work in factories and other forms of wage labour like street selling, which are
particularly characteristic in cities, to unwaged labour in the household and
predominately rural areas, to bonded labour, a form of slavery (Bequele and
Boyden, 1988; Boyden and Holden, 1991). Other activities which are on the
increase include child prostitution, which is prominent in the tourist enclaves
of Southeast Asia in particular. According to the ILO, child labourers are
popular because they will work for long hours for low pay, and are unable or
unwilling to protest against poor conditions.

Current debates centre on whether there is a difference between 'child
labour' (usually waged) and 'child work' (unwaged) in respect of exploitation.
Research has shown that urban children are more likely to work for a wage in
a diverse range of informal occupations, extending from domestic work, rag-
picking, street occupations such as shoe shining and sweet selling, to factory
work (see Plate 8.4). In Jakarta, where environmental legislation dictates that
private cars must carry as least three people within the city's CBD, young
children hire themselves as passengers to drivers entering the city. However, as
Nieuwenhuys (1994) argues, 12 hours of unpaid manual labour on a family
farm can be as detrimental to the development of a child as street selling in
the city.

As a result, current definitions of child labour focus on whether the work is
detrimental to the physical and mental development of the child. Action Aid
(1992), who have developed programmes to provide children with suitable

work and schooling, argue that 'child work', which is often part of normal socialisation, becomes 'child labour' when its character changes from 'developmental' to 'economic'. Other definitions highlight child labour as work which conflicts directly with education, growth and personal development.

Other debates centre on whether the western preoccupation with Third World child workers is further evidence of the enforcement of western codes of conduct on developing nations. Discussions abound as to whether 'childhood' is a contemporary western concept, which is not applicable in the same form to many other nations. Even in Europe, for example, domestic labour was seen as 'ideal training' for working-class girls in past times. Most researchers, however, would agree that although an element of child work may offer a suitable training for children, the exploitation of children in urban employment is highly undesirable in the twentieth century.

Child labour is rooted in poverty, history, culture and global inequality. The fundamental reason why children work is poverty. The incidence and visibility of child labour are more acute in the city due to increasing household vulnerability and the availability of waged work in service and manufacturing industry. Urban industries, frequently export-oriented, which employ children are often hazardous and include the production of carpets, glassware, matches, fireworks, gem-polishing, quarrying and tea-picking. Rag-picking is a common occupation in Bombay, Calcutta, Mexico City, São Paulo and Manila.

The carpet industry is one of the most renowned exploiters of children. It is a boom industry geared towards export, and it operates on a cottage industry basis. 'Dollar Land' in Uttar Pradesh, India, so named due to its profitability, is a region geared to carpet production, where 150,000 children over 10 years old work 10–16-hour days. Anti-Slavery International reports that most of these children have been sold for between £80 and £200 by poor parents. Working conditions are harsh, with poor ventilation and sanitation, which may lead to a range of health problems, including loss of eyesight and lung cancer.

The Indian Constitution bans the employment of children in such work but such legislation is not enforced. Likewise in Nepal, the Children's Act of 1992 prohibits the employment of children under 14 years. In Nepal, it is argued that poverty, lack of child rights and the feudal agricultural system all lie at the root of the problem. Nepal is famous for its 'kamaiya', bonded child labourers who have been sold into bondage in return for small sums or in repayment of debt, particularly to rural landowners (Satteur, 1993). In Kathmandu, children are seen as 'economic refugees', escaping from rural poverty and forming their own street culture. It is argued that governments must address the problems of urban migration by assisting the rural landless.

Child prostitution is growing in the cities of Asia, where in Bombay alone it is estimated that there are 100,000 child prostitutes. Young girls are often taken from impoverished rural areas in Nepal to India on the promise of a better life, although rumours of kidnapping abound. Television programmes and newspaper articles have documented the lives of children working in the carpet factories of Nepal, and the Brazilian and Colombian death squads operating amongst the vast population of street children in São Paulo and Bogotá.

Table 8.5   The United Nations Convention on the Rights of the Child, 1990

Children have the right to:
- enough food, clean water and health care
- an adequate standard of living
- be with their family or those who will care for them best
- protection from all exploitation, physical, mental and sexual abuse
- special protection when exposed to armed conflict
- be protected from all forms of discrimination
- be protected from work that threatens their education, health or development
- special care and training if disabled
- play
- education
- have their own opinions taken into account in decisions which affect their lives
- know what their rights are

Source: Satteur, 1993

There is a growing body of street children in most developing cities. For instance, it is estimated that 2 million children live on the streets in Mexico City. Many of the homeless have run away from rural areas in search of a better life or have escaped from bondage.

As competition over prices and flexibility increases, it is argued that more industries will seek to employ children across the globe. Yet recent GATT negotiations failed to initiate policy directives to halt this trend. Although there is legislation, it is difficult to enforce even in the countries of the North as well as in the South. In 1993, the UN Convention on the Rights of the Child was signed by 100 governments (Table 8.5), but so far appears to have had little effect. NGOs, and other concerned agencies, argue that instead of banning child labour, governments should focus on poverty alleviation and the provision of safer waged work for children. There are now key policies on child labour such as the International Programme on the Elimination of Child Labour (IPEC), operating under the ILO since 1992, which includes a programme of action against bonded labour. Regardless of new legislation, Robertson (1994) argues that whilst free trade abounds, child labour will be commonplace.

Conclusions: work and the city

This chapter has re-emphasised the fact that urban populations are growing faster than the growth of jobs, a trend which is unlikely to change in the near future. As a consequence, developing cities are living with a growing labour force which has fewer formal employment opportunities than previous generations, a situation which is giving rise to escalating unemployment and under-employment. Many job seekers have found productive work in the informal sector, although the ability of the informal sector to continue expanding to meet demand is questionable.

Many practitioners argue that the informal sector, if encouraged through governmental assistance, can be a highly productive engine for growth. Others are more sceptical over whether the informal sector can ever provide more than a 'safety net' or refuge for the urban poor. What is clear, however, is the fact that millions of innovative workers are able to utilise their skills in the informal sector in order to sustain their households, despite pressure from recession and structural economic change. As Chapter 5 has argued, the poor are not passive participants in the quest for empowerment. Through community organisations and new social movements, the urban poor are assisting each other in creating more sustainable livelihoods, using their social capital and networks (Sweetman, 1994). In the 1990s, a recognition of the importance of civil society and governance has promoted co-operation between governments, NGOs and communities in the city. In many cases, the informal sector and its networks have provided a foundation for change.

At the same time, processes of globalisation are extending practices of deregulation and free trade across the globe. Increasingly, developing countries are locations for free trade and export processing industries, which take advantage of cheap female and child labour in an increasingly competitive market. Whilst export-oriented trade does produce new jobs in the South, there is concern over both the quality and value of this new work. Increasingly, informal workers are being contracted by capitalist firms as flexible labourers to work at home in return for a meagre wage.

The dimensions of the employment problem, therefore, go well beyond the shortage of working opportunities for the urban poor, as they are influenced by uneven global development and processes of exchange. An integral part of development is the provision of an alternative employment strategy which identifies three main requirements: the right to work without obstacles as to place and type of work available; a work process which allows for effective worker control; and an end to discrimination based on gender, colour, ethnicity and class. It is not surprising that, given the severe repercussions of recession, the informal sector is gaining favour amongst many social groups. It is also likely that in the present economic and political climate, any move towards a type of work which is more locally attuned, serves the needs of local communities and is less linked to the international economy, may gain momentum. Whether this outcome is likely, given the ever-increasing role of free trade and TNCs in the labour markets of developing cities, remains to be seen.

Chapter 9

Cities and environmental sustainability in the developing world

Introduction

> The children work in poorly ventilated sheds, the air thick with wool fluff, for up to 18 hours a day. Child carpet makers lose their eyesight. Others get problems with their legs from sitting all day in cramped conditions. In the quarries, children get maimed. We are talking about a whole generation being damaged. (Action Aid, 1992: 4)

Recent reports by Amnesty International and the media about the execution of street children in Brazilian cities by police 'death squads' to 'clean up' the urban environment are just one extreme example highlighting the perceptions held by urban municipalities over the causes of environmental degradation and aesthetic blight. In the first half of 1993 alone, Amnesty International reported that 320 homeless children had been executed in Rio de Janeiro in a campaign which regarded homeless children as unwanted 'pests'. In this, and many other instances, it is the vulnerable victims of poverty who are seen by urban governments to be the cause of urban environmental degradation.

Whilst no government would publicly sanction the execution of children as a policy to tackle homelessness, Timberlake and Thompson (1990: 1) argue that this '**child crisis**' results from an unequal economic system which favours rich countries, a situation which is compounded by the concentration of power and capital in the hands of urban elites and multinational corporations. The fact that such atrocities occur is in itself an indication of the vulnerability of certain groups in developing world cities, and it highlights the interrelationships between poverty, the environment and basic human rights.

Paradoxically, during the 1990s the environment was placed at the forefront of the international agenda, particularly with regard to the 1992 Rio Summit (UNCED, 1992). Global institutions, national governments and communities have also become increasingly concerned over the future of the environment, many of these organisations having adopted the concept of '**sustainable development**' (Redclift, 1992). It is apt, therefore, that the penultimate chapter of this book should examine urban environmental conditions, as it is an issue which draws together many of the topics considered in previous chapters.

In its simplest form, there is a fundamental argument which states that a healthy and safe environment is essential for the continued survival of a given population. A population's mental and physical health is directly linked to the quality of the various 'environments' with which its members interact, from the household and workplace to the international scale. Although the figures vary, it has been estimated that approximately 600 million urban dwellers in Africa, Asia and Latin America live in life-threatening environments with respect to overcrowded and inadequate shelter, sanitation and drainage, unsafe housing sites and working conditions, and the absence of primary health care (Hardoy *et al.*, 1992; Kirby *et al.*, 1995). Many urban centres in the developing world have been built on ecologically fragile lands, which are vulnerable to natural hazards such as earthquakes, floods, hurricanes and soil erosion. When natural and human-induced hazards occur together, however, it is usually the urban poor who disproportionately bear the costs.

Although the extent of the interconnection and interdependence between countries is still open to debate, it is now widely accepted that the actions and activities of all nations inevitably affect the lives of others, whether in relation to acid rain, waste generation or consumer preferences. Whilst the need for sustainable development is widely supported, current global economic and social trends are leading to further environmental damage which is not sustainable in the long term. For example, the increased ownership of private cars, the consumption patterns of a growing 'disposable society', and increased materialism, are all incompatible with sustainable development. As developing world countries become wealthier, there is concern that their lifestyles and consumption patterns will start to resemble those of the West. In the 'South', the scale of poverty and the failure of previous decades of development have meant that the majority still lack access to basic needs. This begs the question why developing world cities are still encouraged to develop through economic growth when western consumption and resource use patterns are so evidently non-sustainable. If all the poorer nations of the world were to consume the same amount of resources as the United States, our planet would not be able to sustain life for long. On the other hand, should poorer nations be encouraged not to 'develop' in this way, so that the richer countries of the 'North' can continue their exploitation of the natural and human resource base? As the Rio Summit highlighted, developing countries have found their ability to react to global environmental problems constrained by their concern over the daily survival of their populations.

As this chapter will show, there are many contradictions between the international call for sustainable development and current global economic trends. As Chapter 8 illustrated, neo-liberalism and the New International Division of Labour are leading to increased privatisation of public resources, the exploitation of workers through deregulation of rights and safety conditions, and the spread of 'footloose' multinationals which are able to avoid local environmental legislation in countries without the finance and means to enforce tighter controls. Increased privatisation and deregulation is also occurring at a time when the state needs to be centrally involved in environmental planning (Hardoy *et al.*, 1992; Main and Williams, 1994).

Through an analysis of the concept of '**sustainable development**' and current debates on the '**sustainable city**', this chapter argues that a healthy and secure environment is a prerequisite for successful urban development. As the environment is simultaneously the cause and a victim of urban problems, it is an issue which links together shelter, employment, basic needs, poverty, human rights, politics and gender. The chapter moves on to examine the World Bank's **Brown Agenda** and the way in which environmental issues are gendered, before examining a number of key environmental priorities for developing world cities, namely adequate access to safe and secure shelter, basic services and working environments. Before concluding, the chapter argues that urban environmental sustainability is ultimately about **empowerment**, **good governance** and **legislation**, and examines the positive role of local environmental movements and community action groups in working towards sustainable cities through the distribution of power and participatory policies. After addressing the question 'who is best suited to manage the urban environment?', the chapter concludes that poverty alleviation, environmental sustainability and the empowerment of the poor need to be addressed simultaneously, in order for any real change to occur in the immediate future.

Sustainable development and cities in the developing world _____

The 'sustainable development' debate

The concept of sustainable development has dominated both the environmental and development literatures since the early 1980s, when decades of mismanaged growth in the name of 'development' were manifested in environmental degradation at a global scale (Redclift, 1987, 1992; Simon, 1989b; World Conservation Strategy, 1991; Adams, 1990; Shiva, 1992). 'Sustainable development' is a popular catch-all phrase which endeavours to encapsulate growing concerns over the future of the planet by highlighting the inextricable links between environment and development (Mannion, 1992). Despite its global popularity, there is no universally agreed definition of sustainable development, which according to Redclift (1992, 1994b) is precisely because the term expresses different views of 'development' itself (see also Kirby *et al.*, 1995; Holmberg and Sandbrook, 1992).

Since the Brundtland Commission published its report *Our Common Future* in 1987, the sustainability debate has revealed major differences in thinking about development, economic growth, social change and environmental conservation (World Commission on Environment and Development (WCED), 1987; Redclift, 1994a). For example, environmentalists who argue from a 'green' perspective identify the natural resource base as the focus of the sustainability debate (Rees, 1990; Lohmann, 1993) which must be protected at all costs. Other writers take the view that it is more important to consider sustaining present and projected future levels of production and consumption in order to enhance human development. In this respect, sustainable development can imply a radical change in lifestyle to manage existing resources in a more sustainable manner (Pearce *et al.*, 1989). For example, Munasinghe (1993: 16) argues that the economic goal is 'to increase the net welfare of economic activities while maintaining or increasing the stock of economic, ecological, and socio-cultural assets over time'.

Whilst advanced capitalist states in the 'North' might be more concerned with the natural environment and conservation, the sustainable development debate in the 'South' often prioritises human development, which as Barbier (1989: 103) argues 'is directly concerned with increasing the material standard of living of the poor at the "grassroots" level'. The emphasis here is on meeting **social and economic objectives**, with a focus on improving access to basic needs (see Chapter 5). In particular, there is a need to ensure that the poor have access to sustainable and secure livelihoods, as it is the poor that often have no option but to choose short-term economic benefits at the expense of the environment.

These contradictions between the views of the 'North' and 'South' require attention to be paid to the structural inequalities of the global system (WCED, 1987; Johnston *et al.*, 1996; Kirby *et al.*, 1995). In developing areas, environmental struggles are often about basic needs rather than enhancing already comfortable lifestyles. This takes us to the Brundtland Commission's definition of sustainable development, where the basic emphasis was on meeting human needs: 'development which meets the needs of the present, without compromising the ability of future generations to meet their own needs' (WCED, 1987). This is perhaps the most widely used definition by both academics and practitioners alike, but one which is open to a variety of different interpretations.

According to WCED (1987), the main components of sustainability are **environment**, **equity** and **growth**, although some practitioners would question the role of growth. Although the notion that human development needs, including cultural and social ones, must go hand in hand with long-term environmental considerations has gained much support over the last decade, there is still a fundamental problem concerning the global agenda for sustainable development. The 1992 Rio Summit highlighted the lack of consensus over the sustainability debate between the 'North' and the 'South', particularly with respect to different environmental priorities as well as concerns over the financing of this new agenda (UNCED, 1992; Middleton *et al.*, 1993; Hardoy and Satterthwaite, 1995; Redclift and Sage, 1994).

Essentially, the 'North' failed to appreciate the need for a global approach to the problem, as it insisted on greater efforts by the 'South' to conserve essential resources, such as rainforests, without giving consideration to the poverty of many countries and the financial constraints imposed by the debt crisis. In their analysis of the Rio Summit, Kirby, O'Keefe and Timberlake (1995: 10) argue that 'the north turned "green" and the south was turned away'. For countries in the 'South', long-term environmental objectives are difficult to comprehend when short-term problems pose the most serious risks. Despite the heightened awareness of environmental factors, there has been no move to assist poorer nations to reduce their debt burden, thereby enabling them to reduce exploitation of their natural and human resources for foreign exchange (George, 1988).

The levels of environmental degradation found in many developing world cities clearly indicate that the global system and the trend towards westernised living are far from sustainable, and on the whole are particularly damaging. However, an immediate agenda for radical change has so far been obscured by failing to face the fact that satisfying basic needs is a necessary step to ensuring the viability of **Agenda 21** (UNCED, 1992). The environment has to an extent become the unacknowledged victim of inequitable development and political power (George, 1988).

Sustainability and 'development': which way now?

It is also important to appreciate how the concept of sustainable development is intrinsically linked to current arguments which revolve around the concept of 'development' itself. Attempts to 'westernise' the Third World through Eurocentric economic and industrial strategies largely contradict the ethos of sustainability (Shiva, 1988; Mehmet, 1995). Why, for example, is economic growth still advocated by many development agencies as the principal way forward, when the lessons of history suggest that this is often a mixed blessing? As western-styled progress is centred on growth, most development strategies will by definition be unreconcilable with environmental sustainability. As the main problem rests on the fact that human progress almost invariably affects the environment, many authors now argue that we should be questioning what development means, as well as how it happens (Mies and Shiva, 1993; Norgaard, 1994).

Continued encouragement by the World Bank of growth, greater urban productivity, and the adoption of western technology is pushing societies further away from traditional, more sustainable forms of 'development' (Mies and Shiva, 1993). This premise is based on the argument that small-scale societies and economic activities are closer to the natural environment. Regarding them retrospectively, many of the development programmes concerned with modernisation have resulted in increased poverty and environmental degradation (Braidotti et al., 1994). In her discussion of the 'myth of catching-up development' Mies (1993) argues that poorer countries should follow an alternative path based on subsistence and community development. It is important, however, to recognise that any kind of 'development', including access to basic needs, will require some increased use of natural resources (Hardoy et al., 1992). It is a fact that those developing world cities which use least environmental resources, many of which are to be found in Africa, are those which lack basic services. Drakakis-Smith (1995: 662) argues that for many authors, '"sustainable growth" is a contradiction in terms since "sustainable" relates to the placing of limits on growth'. Following this line of argument, a more realistic working definition for developing world cities might be '**sustained development**' within the context of continual growth (see also Douglas, 1983, 1989; Cernea, 1993). Indeed, are there any alternatives at present, when cities are continuing to grow? In many poorer countries, improving the quality of, and access to, basic needs will require a continued or increased use of natural resources, particularly in the context of the built environment.

The sustainable city

There has been much work on the concept of the '**sustainable city**' with respect to cities in the developed world. In Europe and North America, governments and planners are promoting the **compact city**, which favours high-density living and encourages populations to move back to inner cities, thereby reducing travel-to-work times and energy consumption. As Chapter 6 highlighted, there is little evidence to suggest that cities in the developing world are undergoing similar changes. Moreover, in many developing world cities, the main environmental concerns are not energy reduction or the ozone layer but rather more immediate concerns over water, sanitation, shelter and access to secure livelihoods. It has been argued by Hardoy, Mitlin and Satterthwaite (1992) that the growing interest in urban environmental problems is based on the assumption that environmental problems in the developing world are the same as those in the developed world.

Far less consideration has been given to the role of developing world cities in the sustainability debate than is warranted, although there is a growing body of literature which examines the relationship between the environment and urbanisation (Hardoy and Satterthwaite, 1989, 1995; Satterthwaite, 1992; Main and Williams, 1994; Elliott, 1994; Stren *et al.*, 1992). There are a number of key debates which are currently being addressed.

Firstly, given the lack of consensus over sustainable development, it is not surprising that there is a similar debate over the true meaning of the 'sustainable city', particularly in relation to its national economic and social role. There has been a call for an increased understanding of what is meant by sustainable urban development, so that issues can be prioritised and appropriate policies formulated (Goodman and Redclift, 1991; Drakakis-Smith, 1995). In his informed discussion of sustainable cities, Satterthwaite (1992: 3) argues the need to search for 'cities (and rural areas) where the inhabitants' development needs are met without imposing unsustainable demands on local or global natural resources and systems'. He further argues that it is not cities themselves which pose a threat to sustainability, but the production and consumption patterns associated with them, particularly those of wealthier income groups. Therefore, it is ineffective to conceptualise sustainable development solely in terms of 'environmental' objectives, when it is also the political and economic arena which fails in its management and control over these resources (see Chapter 5). As will be discussed later in this chapter, it is imperative that access to, and command of, resources for low-income groups is seen as a major objective of sustainable urban development.

Secondly, there is a lack of consensus as to whether developing world cities should be seen as centres of economic productivity as proposed by the World Bank, or centres of environmental problems and concentrations of social deprivation (UN, 1995), or whether they are locales from which to promote equitable and sustainable development. Thirdly, Kasada and Parnell (1993) have highlighted the importance of addressing the relationship between urban size, prosperity and environmental problems. Although smaller cities may find growth more difficult to accommodate, an increase in scale does not necessarily lead to a corresponding rise in environmental problems. For instance, although Curitiba is one of Brazil's fastest growing cities, it has made substantial improvements in the quality of life of its residents through an innovative public transport system, the integration of social programmes and environmental education, and the preservation of the city's cultural heritage (Rabinovitch, 1992). Rather, environmental problems become more serious when the urban population increases without the appropriate institutional framework.

Whilst the reasons for the existence of insanitary conditions and environmental degradation in developing world cities are frequently seen as a lack of resources and poverty, there is little evidence to suggest that increased economic growth translates into increased environmental concern and quality of life for urban residents. In fact, many of the cities with the worst environmental problems are located in the 'miracle' economies of Pacific Asia. For instance, with 6 million inhabitants, Hong Kong produces 23,300 tonnes of solid waste and 21 tonnes of floating refuse, 2 million tonnes of sewerage and industrial waste per day and 100,000 tonnes of chemical waste per year (Chan, 1994).

Fourthly, following recent work undertaken by McGee and Yeung (1993), attention has been paid to the problems associated with 'extended metropolitan regions' (EMRs), where there is a blurring of rural–urban functions. These regions place huge pressures on precious resources such as water and energy, particularly in relation to competition between domestic and manufacturing use. As many authors state that the growth of EMRs is the 'inevitable future of urbanisation' (Drakakis-Smith, 1995: 661), city legislation and planning are likely to become more difficult as municipal boundaries are blurred. As White and Whitney (1992) argue, many cities have already exceeded their carrying capacity and the consequences are usually expressed on their margins.

It can be argued that cities should be better placed than rural regions to provide healthy, safe environments which meet sustainable goals (Potter, 1996). These goals include a healthy living and working environment, safe water supply, the provision of sanitation, drainage and garbage treatment, paved roads, an adequate economic base and good governance (Hardoy et al., 1992). Inherently, this requires the maintenance of a sustainable relationship between the demands of consumers, institutions and firms, with resources and ecosystems. There are, however, a number of other important social and cultural goals which will improve city living for the majority. These include the maintenance of social institutions, public spaces and amenities, and the representation of local culture in the form of the built environment. Ultimately, the state of the environment and the lives of people in the city depend also on the political structure and the behaviour of planners and municipalities, as well as the national government.

Meeting human and environmental needs in the city: major issues

The Brown Agenda and the gendering of environmental issues

The World Bank's **Brown Agenda** (1991, 1993b) attempts to prioritise a range of environmental problems which are currently facing developing world cities. Although the agenda focuses on well-established factors such as population growth, basic needs and economic productivity, it stresses the need for local and national governments to control growth and enforce environmental legislation in order to reduce people's vulnerability to environmental hazards (Leitmann, *et al.*, 1992). Despite its good intentions, the agenda has been criticised for failing to recognise the importance of politics and power in governing the poor's access to urban services (Ferguson and Maurer, 1996; Mueller, 1995).

As previously indicated, the consequences of environmental deterioration disproportionately affect the poor. As this section will show, the poor are often forced to build shelters on high lands and swamps or near polluting industries without access to water and sanitation, and are more likely to work in hazardous occupations with no health and safety legislation. As Chapter 5 discussed, many developing world cities are characterised by high rates of infant mortality, limited life expectancy, disability and work-related injury, disease, poor quality of life and mental health. These environmental issues are often gendered, as it is often women who are directly affected by environmental problems and have correspondingly organised to promote change. Unfortunately, strategies like the Brown Agenda have not really addressed the gendering of the environment.

The influence of **gender relations** in the human use of the environment is of paramount importance in many developing countries (Braidotti *et al.*, 1994; Jackson, 1993; Shiva, 1988; Harcourt, 1994b). There is now a growing recognition that the different socio-economic roles played by men and women have profound environmental implications. As Chapter 5 outlined, women often undertake a primary role in the care of the environment, and are more likely to be affected by environmental problems, particularly those related to the home, workplace and health. Whilst appreciating the fact that women from different societies and backgrounds do not represent a homogeneous group, it is still possible to outline a number of common relationships which show how a gendered approach to environmental issues is of paramount importance in developing world cities.

In most cities, the main victims of environmental degradation are the poor and vulnerable, and women constitute a large proportion of this group. As Agarwal (1992) notes, women are at once both victims and agents of change. Despite increasing awareness amongst international organisations, women's involvement in, and control over, natural resources is still undervalued. At the end of the United Nations Decade for Women in 1985, there was a call for increased attention to be paid to women's links with the environment

(Dankleman and Davidson, 1988). This has been met by women's local environmental groups rather than by global institutions. All too often, development agencies still implement projects which marginalise women's knowledge and experience of their local environments. In order to appreciate the importance of undertaking a gendered approach to urban environmental policy, there is a need to understand the arguments which identify women as the key actors in sustainable urban environment.

It has been argued from a variety of environmental–feminist approaches that women and nature are closely linked together. Firstly, it has been argued that due to their role as givers of life, women 'naturally' care more than men about the environment (McCormack, 1980). Although this line of thought has gained much support due to positive notions of women as the more harmonious gender, it has also been criticised as maintaining traditional stereotypes which identify women as more 'caring', 'gentle' and 'passive' due to their reproductive role (see Braidotti *et al.*, 1994; Harcourt, 1994a). Do women naturally care more about the environment, or is it the gender division of labour that places them disproportionately in this role?

Secondly, some feminists link women and nature together as they are both seen as the 'other', and have been dominated by external interests for so long (Mies and Shiva, 1993). Shiva (1988) argued that environmental degradation has resulted from a pattern of '**patriarchal maldevelopment**' associated with capitalism, a line of thought which ignores the areas where women have been involved in environmental degradation themselves. From an '**ecofeminist**' perspective, Mies and Shiva (1993) argue the need for a 'new vision' which promotes greater self-reliance and environmental sustainability through new forms of economic activity based on more traditional and informal work, grassroots democracy, common responsibility and subsistence technology.

Thirdly, women play key roles as invisible managers and end users of the environment in many cities, particularly in respect of natural resources such as water, land and energy (Sontheimer, 1991). In the urban context, women's involvement and control over water and sanitation is a crucial issue for sustainable development, as is their access to land for shelter (Moser *et al.*, 1991; Chant, 1996b). For instance, women are often responsible for water collection and purification, and the health and hygiene of the household and the community. As will be discussed in the next section, they also undertake major roles in hazardous commercial and multinational enterprises, and are more likely to utilise their homes as informal workplaces, thus raising implications for health and safety (Moser, 1995a). As the reviews of the living and working environments facing urban residents will show, it is women who are best placed to utilise their local knowledge to promote sustainable urban development. The rise of women's urban action groups in order to fight for basic services, shelter and work has shown their capacity as 'innovators' of urban change.

The living environment: shelter and basic services

Within developing world cities, access to safe land, adequate shelter and basic services constitute fundamental human rights which must underlie any notion of sustainable development (Hardoy and Satterthwaite, 1989; Hardoy et al., 1992; Satterthwaite, 1995). As previous chapters have illustrated, the increasing pressure placed on cities in the form of growing populations, unequal control of, and access to, resources, and poverty, has culminated in the degradation of the living environments of many urban citizens. Potter (1992b, 1996) has argued that more than 50 per cent of the world's population live in substandard shelter, a factor which has a direct effect on their quality of life and health. Douglass (1992) predicted that 60 per cent of the urban population of Asia would be living in slums or squatter settlements by 2000. Although the nature and scale of environmental problems differ from city to city, most developing world cities have expanded without a corresponding increase in services and appropriate planning, and with no control over environmental degradation or a suitable institutional framework for urban governance. As a result, a number of negative environmental trends have come to light.

Firstly, as a result of uncontrolled growth, changing land uses and urban development, low-income housing has increasingly spread into marginal areas which are vulnerable to physical and human-induced hazards. Although a high proportion of Third World cities are located in areas vulnerable to natural hazards, such as earthquakes and hurricanes, most disasters are in fact human-induced. Of particular concern is the spread of informal settlements on unstable marginal land which endangers human life (Potter, 1996). Of critical importance here is the unequal nature of urban land markets, and the restricted access by the poor to safe and legal housing, as already discussed in Chapter 7. As Dankleman and Davidson (1995) have argued, over half the inhabitants of some cities have either settled on vacant lots or live as squatters in slums.

In the last decade, many low-income communities have been destroyed by landslides, flooding and earthquakes. Main (1994) documents the occurrence of at least 14 urban environmental disasters arising from severe weather in the Third World between February 1992 and August 1993, and a further nine human-induced disasters between January 1991 and August 1993. Degg (1994) estimates that 78 per cent of the developing world's largest cities are exposed to one or more natural hazards, including earthquakes, volcanic activity and tsunamis. Earthquakes, in particular, present a serious hazard to most large cities. For instance, the Mexico City earthquake of 1985 led to the death of over 5000 people, mostly as a result of slope failures in the central *barrios* (Degg, 1994), and the landslides in Rio de Janeiro in 1988 were more about uncontrolled *favela* development than natural vulnerability.

Main and Williams (1994) argue that there are many 'negative externalities' which increase with the scale of urban development. In Caracas, Jimenez-Diaz (1994) states that in 1985, 61 per cent of the total population lived in *barrios*, and that 67 per cent of the land under *barrio* construction was unsafe enough to warrant eviction. It is further reported that the incidence of landslides in

Caracas has increased from one per year in the 1950s and 1960s, to an average of 25 between 1971 and 1979, to 33 per year during the period from 1980 to 1987. Whilst the earlier landslides were associated with seismic activity, uncontrolled *barrio* development has exacerbated their frequency. In the 1974–79 period, the main areas of landslide activity were areas of dense *barrio* development. When tropical storm Bret hit Caracas on 8 August 1993, at least 150 people were killed and thousands lost their homes as a result of mudslides in the hillside shanty towns (Potter, 1996). A similar association between increasing informal sector housing activities and increasing slope failures has recently been documented in the Niteroi Municipality of Rio de Janeiro by Smyth (1996).

Elsewhere, low-income communities have been forced to build homes on flood plains, swamps or coastal waters. When hurricanes swept the Caribbean in the summer of 1995, the island of Trinidad was affected by strong winds. However, severe damage occurred in Sea Lots, Port of Spain, an area of squatter housing located around the port. Potter (1996) states that such hazard vulnerability is as apparent in small cities of the Caribbean such as Kingstown, the capital of St Vincent and the Grenadines, as it is in the mega cities of Rio de Janeiro and Caracas. In African cities, flooding is often the greatest environmental hazard. In August 1988, heavy rains washed away refugee settlements in Khartoum, affecting around 800,000 people.

Secondly, urban environments frequently suffer from a high incidence of **biological pathogens** and disease resulting from restricted access to water and sanitation, which is compounded by overcrowded and cramped living conditions, and poor waste disposal. As discussed in Chapter 5, the WHO Commission on Health and the Environment in 1992 cited the incidence of biological pathogens and lack of access to safe water as the most common cause of illness and premature death in Third World countries. A safe water supply would dramatically reduce the incidence of diarrhoea, dysentery, typhoid, intestinal infections and cholera (Cairncross *et al.*, 1990; WHO, 1992).

Many urban residents still obtain water from standpipes, polluted rivers and water vendors. Hardoy, Mitlin and Satterthwaite (1992) estimate that between 20 and 30 per cent of the urban population in the developing world obtain water from vendors. For example, most Southeast Asian cities still lack safe water and sewerage disposal systems (Dwyer, 1992). In Jakarta, 40 per cent of households are dependent on water vendors and have no water-borne sewerage system. Only 15 per cent of the population have access to the city's water supply. In Manila, the sewerage system built in 1909 for 440,000 residents is now the only facility for 6 million people. In Karachi, only 38 per cent have piped water, 46 per cent use standpipes and 16 per cent rely on vendors (Hardoy *et al.*, 1992). Smaller cities in Africa are often the worst serviced, as in Dakar, where there is no provision for the removal of waste. Inappropriate technology, like the construction of expensive flush toilets, is frequently employed by external agencies to solve sanitation problems.

In poor city areas in the developing world, infants are 40 times more likely to die before their first birthday than those in Europe and North America, mostly as a result of environment-related causes (Cairncross *et al.*, 1990). Over-crowding, estimated at an average of over four persons to a room in developing world cities, leads to widespread bacterial and viral infections. Cramped cooking facilities, and pollution from charcoal fires and kerosene stoves, are particularly hazardous to women and children. The multi-functional use of the home and community for urban production also presents another set of environmental externalities, an issue which is considered in the next section.

The work environment: access to secure livelihoods and industrial development

> Cross over the border bridge from the wide, neat streets of Brownsville to Matamoros and drive for 10 minutes to the nearest colonia – and the hard-won workers' rights of the past century fall, literally by the wayside. Factories openly dump huge mounds of calcium sulphate residues on to street pavements; children play in pools of toxic green scum and treacly black chemical wastes are dumped into ditches within yards of workers' shacks. (Ghazi, 1993: 18)

The Urban Foundation (1993) considers **increased economic production** to be the main way in which developing world cities can finance urban environmental improvements. Similarly, the World Bank (1991), along with other global institutions such as the ILO (1995), have also identified urban productivity as a major strategy for alleviating poverty. This agenda, which often focuses on the **formalisation of the informal sector**, also encourages industrial growth and resource depletion. As shown in Chapters 3 and 8, developing countries are often persuaded to increase their economic production through foreign direct investment and multinational export production. If, as it appears, the World Bank and 'Northern' governments believe that this should take place in a privatised 'free market' economy, what power will the state have in providing a suitable legislative framework? This is of particular concern when current global trade arrangements which favour the 'North' will further reduce the ability of the 'South' to sustain its own environments. These fundamental contradictions between philosophy and policy highlight the enormous obstacles which need to be overcome if real progress towards sustainability is to be made.

Most governments in the developing world have legislation designed to govern work practices, such as health and safety, as well as to regulate industrial environmental degradation. The main problems are twofold: firstly, policy instruments are often ineffective and not enforced; and secondly, unregulated market forces and the quest for profit maximisation by firms invariably result in unregulated construction, the unsuitable location of industry, the exploitation of human resources and widespread pollution. The appalling realities of the lives of Third World workers who are supposed to enhance 'urban productivity' must be understood by those institutions who advocate unorganised labour practices as a development strategy.

Developing countries now account for 13.2 per cent of global industrial production, partially resulting from the movement of heavily polluting industries from the 'North' to the 'South' (Robins and Trisoglio, 1995). The NICs have been the recipients of much of this industry (textiles, petrochemicals, iron and steel), as in Thailand, whose share of hazardous waste-generating industry rose from 29 per cent in 1979 to 58 per cent in 1989 (Phantumvanit and Panayotou, 1990). Tighter environmental control in the United States has prompted the movement of its asbestos industry to Brazil and Mexico. This situation is potentially dangerous as few countries in the 'South' have the legislative structure and technology to deal with hazardous waste.

The media has brought worldwide attention to the role of multinational enterprises in environmental and human disasters in the last two decades. Due to failures in planning regulations, and the encroachment of low-income settlements into marginal lands, hazardous industries have been located in areas of high population density. It was in 1984 that the world was witness to an alarming number of industrial-related disasters which affected poor communities. The most infamous example was the explosion at the Union Carbide factory in Bhopal, India, when 3300 died and around 150,000 people were injured due to the close proximity of two large slum developments (Jones, 1988; Jaising and Sathyamala, 1995). Ten years later, reports of the continued ill health and suffering from chronic lung diseases of survivors and their children are largely ignored (*Guardian*, 1994). Also in 1984, a PEMEX (nationalised petroleum company) factory in the San Juan Ixhualepec district of Mexico City exploded, killing 500 people immediately, injuring 7231 nearby residents and leaving 39,000 homeless (see Walker, 1994). Since the factory was built in the 1960s, dense low-income housing had developed along the southern boundary of the complex. It was even reported that families were living amongst the gas tanks in the plant itself. Claims for compensation, particularly as a result of losing essential wage earners, in both incidents were extremely prolonged and highly disputed by the companies involved.

Other more recent examples of unregulated urban development include a number of factory fires which have resulted in large losses of life. Female workers in export factories producing goods for foreign markets appear to be particularly vulnerable as an industrial fire in Bangkok illustrated. On 12 May 1993, a fire swept through a soft toy factory located in an industrial zone on the outskirts of Bangkok (*Guardian*, 1993). Despite the occurrence of two other fires in the preceding months, there were no fire alarms or fire escapes for the 1800 workers employed in the four-storey building. Although 200 workers died and several hundred were injured, this incident failed to instigate any improvement in the safety of factories. As Potter (1996) highlighted, this incident was followed by a number of similar fires in China's special factory zones.

Multinational, and also nationalised, industries regularly inflict severe environmental abuse on local communities, although this goes largely unrecognised. From the dumping of toxic wastes to the continued use of chemicals such as DDT, which have been banned in the 'North', many companies involved in manufacturing goods for foreign markets do not pay for their negative externalities. Chemical and textile industries, in particular, have blatantly discharged toxic effluents into nearby rivers or released gases into the atmosphere. Some wastes, such as lead, arsenic and cyanide, are poisonous and carcinogenic, whilst others carry bacteria and disease. Weir (1995) gives an example of the multinational chemical giant Bayer, which is located in the Belford Roxo district 20 miles from downtown Rio de Janeiro. Low-income communities which are situated downwind of this complex constantly suffer from pollution, toxic smells and a 'reddening' of the local river which periodically floods homes. Similarly, the Kalu River, which runs through Bombay's industrial suburbs, receives liquid effluents carrying lead and mercury from 150 industrial companies (Hardoy and Satterthwaite, 1995). For many poor urban communities, rivers such as these are their only source of water, and they are also areas where children play and work.

It is the *maquiladoras*, located along the Mexican-US border, where some of the most atrocious examples of industrial pollution and environmental damage by multinational companies have occurred (Sklair, 1994a). As discussed in Chapter 8, industries are attracted to the Mexican border by cheap wages and lax environmental policy. It is estimated that around 700 *maquiladoras*, including Fisher-Price and General Motors, are located in the export-processing city of Tijuana which employs 350,000 workers. Tijuana and the Rio Grande have become notorious dumping grounds for pesticides, battery acid and toxic industrial waste by multinationals. *Agua negra* (black water) is said to flood streets and homes, rates of cholera and cancers are reported to be high, and children have been born with brain and kidney damage (Ghazi, 1993). *Maquilas* are Mexico's second-biggest source of foreign exchange following oil. In Matamoros, a Mexican border settlement of 500,000 inhabitants which is home to 2100 *maquila* industries, workers suffer from illnesses associated with toxic chemicals, inadequate lighting, space and ventilation. The Mexican government, although attempting to change the operations of companies, is acutely aware of the fact that it needs this foreign investment at all costs. In 1992, Presidents Bush and Salinas signed the Integrated Border Agreement, under which the United States and Mexico would try to clean up the region in preparation for further expansion. Since then, companies have been asked voluntarily to reduce toxic emissions and a number of companies have been temporarily shut down. In these industries, workers as well as the natural environment also suffer from unlawful practices.

Globally, there are 32.7 million occupational injuries and 146,000 deaths per annum, with a large proportion occurring in the developing world. As discussed in Chapter 8, many workers in export industries, such as the *maquiladoras*, have no sick pay, disability benefits and health care. Employees, many of whom are women, work long hours without rest periods or access to sanitary facilities, and return at night to cramped dormitories. The highest levels of risk are found in asbestos, chemical, cement, glass, iron and steel, mining, textile and leather industries, the same industries which frequently employ children. As Chapter 8 highlighted, prevalent environmentally related occupational diseases include silicosis, blindness, tuberculosis and hearing loss. In Kanpur, at the heart of India's industrial region, 60 per cent of children are reported to have tuberculosis as a result of insanitary working conditions. Their machinery is often old and dangerous, as well as being operated without safety equipment or protection.

As previously discussed, poor households frequently utilise their homes for productive activities, and the environmental sustainability of the informal sector is open to great debate. In some ways, the informal sector is seen to be a more sustainable form of production as its small-scale activities utilise traditional technology and local resources. Many urban dwellers make their living from rubbish dumps, recycling non-renewable materials such as rags, glass and metal (Furedy and Alamgir, 1992). In Delhi, for example, the ILO operated a project which employed around 3000 women in the recycling of old shoes (Prasad and Furedy, 1992). In the Caribbean, informal traders collect old *Carib* (beer) bottles and reuse them for selling channa (chickpeas), nuts and processed goods (Lloyd-Evans and Potter, 1996). As noted in the introduction of the chapter, however, many governments periodically 'clean up their streets' by jailing informal street traders (Stren *et al.*, 1992).

On the other hand, due to its uncontrolled and unregulated nature, the informal sector can also provide a haven of exploitation and environmental hazards. Small informal industries which utilise poorly maintained machinery and produce toxic wastes are frequently located at home. The manufacturing of small parts for vehicles and mechanical servicing, for example, often takes place in residential houses and communal yards with no facilities for waste disposal. Furthermore, subcontraction has led to an increase in home-based production, where women work in ill-equipped environments and frequently handle toxic glues or lead-based chemicals. Thus, the responsibility for health and safety is all too often transferred from the employer to the employee and the community. Despite these important concerns, however, the operation of the informal sector does illustrate the innovative capacity of individuals and their communities to organise for their own development, and our concluding remarks stress this.

Conclusions: empowerment and good governance in the city _____

It is coming to be appreciated that the failure to develop **good governance** underlies many environmental problems in developing world cities. Whilst priority has been given to the global agenda through the development of environmental institutions, action has been most successful at the local scale. In particular, there is a need for alternative avenues of political control which improve the capacity of the poor to manage their own environment. It is the most vulnerable in the city who are adversely affected by environmental degradation. Although poor communities may have no choice but to exploit natural resources for their own daily survival, they often have the knowledge and experience to promote a more sustainable way of living. Douglass (1992) argues that sustainable development must empower households and communities to participate as active decision-makers in the urban environment. As Hardoy and Satterthwaite (1990: 237) have argued, 'the achievement of an alternative urban future depends on the extent to which poorer groups are able (or allowed) to organise not only within their district but also to become a greater political force within the city and the nation'.

Although new social movements may be a truly viable alternative to auto-cratic top-down state policies, there is a danger of placing too great a burden on community organisations as the only **agents of change**, an argument which parallels that in relation to self-help housing in Chapter 7. It is imperative that national and municipal institutions are also engaged in this agenda. Third World NGOs can play a crucial role in strengthening the relationship between communities and local government to ensure a redistribution of decision-making functions and resources at the grassroots level. This is a difficult task as each city will need to structure its own agenda (Satterthwaite, 1995).

In order to protect the environment for future generations there is need for a reorientation of societies and economies towards more sustainable practices, a task which requires greater co-operation between grassroots organisations and international institutions if it is to be successful. Unfortunately, there is all too little support for low-income organisations, when their capacity to work effectively and efficiently has frequently been proved. Many of the most suc-cessful environmental movements are women's groups who have organised for better urban services (Moser, 1995b). Braidotti et al. (1994) argue that NGOs have formed a partnership with women from the South to shift percep-tions of women from 'victims' of the environmental crisis to 'managers' of the environment.

The state of the urban environment is crucial to any discussion of future urban policy, including strategies for poverty alleviation. Much of this volume has highlighted the importance of the various actors such as the state, the private sector, community organisations and non-governmental organisations in working towards a new environmental agenda. Such debates, however, have highlighted the fact that there is often a mismatch between global and local objectives, between the 'North' and the 'South', as well as within cities in any one country. At the macro scale, there is a pressing need to consider equity, social justice and human rights, access to basic needs and greater environ-mental awareness. At the community level, priority must be given to employ-ment, shelter and poverty alleviation, as poverty undermines households' ability to control their own sustainable development (Schwarz, 1993; Peil, 1994). An improvement in the daily lives of many households depends on all these factors being met simultaneously: a tall order, but one that must be met.

Chapter 10

The future of the city in the developing world: the policy agenda

Massive urban growth and development are ongoing, inevitable processes in the developing world. Whilst the population of major cities may have been expanding at a lower rate since the 1980s, especially in Latin America and the Middle East, it is certain that urban areas will continue to grow massively, especially cities of intermediate and secondary size. In addition, cities in South Asia and Africa are still growing at very rapid rates, and as exemplified in Chapter 1, there is little ground for complacency.

The same is true with respect to regional inequalities in developing countries, and also in connection with differences between individual nations making up the global system. Whilst the World Bank currently maintains that gross disparities between the countries of the South and the North are at long last beginning to narrow, just as the gulf between primate and secondary cities is narrowing in Latin America (Gilbert, 1993a), Broad and Landi (1996) have shown that it is the influence of the relatively wealthy NICs that is creating such a rosy picture. At the global level, it has been argued that if the NICs are excluded from the picture, then North–South differentials actually appear to be increasing. Indeed, Broad and Landi (1996) suggest that the North–South gap continues to widen in all but a dozen or so Third World countries, and that there remains a net flow of resources from most Southern nations to the North. This parallels the argument presented in Chapter 3 concerning global proportions of manufacturing in the developing countries declining outside the Newly Industrialising Countries.

Some of these trends are made clearer if we look at actual conditions in Third World cities. World Bank data show that overall levels of poverty increased in most parts of Africa and Latin America during the 1980s, whereas poverty showed reductions in most parts of Asia during the same period (World Bank, 1990; Gilbert, 1992). Thus, data from the United Nations (1989a) indicate that whilst the proportion of people classified as living in poverty in the developing world as a whole fell from 52 per cent to 44 per cent between 1970 and 1985, over the same period those living in poverty increased from 46 to 49 per cent in the case of Africa. But it should be emphasised that at the global scale, the number living in conditions of poverty increased in absolute terms from 0.94 billion to 1.16 billion between these two dates. Further, the scale of urban poverty has been greatly underestimated in the recent past, and grew rapidly in the 1980s (UNCHS, 1996).

Since 1980, cities in the developing world have been hard hit by world recession, the debt crisis and general economic decline. Africa, Latin America and the Caribbean have been particularly affected (Gilbert, 1993a). International agencies and governments have followed the ideological policies of the New Right and have introduced austerity measures. As discussed in several chapters in this volume, policies of privatisation have been implemented, accompanied by the rolling back of the state. Reaganism and Thatcherism have promoted the wider view that the market knows best, and that the poor should be encouraged to help themselves. The main plank of such an approach has been the structural adjustment programme (SAP), involving moves towards privatisation and the promotion of market-oriented systems, public sector cutbacks, retrenchment, reduced wages, the removal of subsidies and devaluation. Gilbert (1993a) has argued that such policies have in some instances started to achieve what governments had not managed to do by means of regional policies, that is to reduce the rate of growth of a number of large metropolitan areas. By means of such approaches it is hoped that in the long run, sustained economic growth will be the outcome. However, these policies are frequently associated with rising levels of urban poverty, due to unemployment and wage reductions coupled with poor access to adequate health care, schools and other basic services (see, for example, La Guerre, 1993; Simon *et al.*, 1995; Gibbon, 1996).

It has been made clear throughout this text that thinking about Third World urbanisation and the major objectives of intervention have changed several times during the twentieth century, as indeed have attitudes to the desirability of intervening in the first place. Firstly, it has come to be accepted that urbanisation in developing countries is inevitable (McGee, 1994), and is not something which can be cleared away or hidden from view. As explored in Chapter 2, increasingly the view is being taken that urban policy must be looked at in the context of development policy as a whole. But since the 1980s, the major change has, of course, been the inexorable drift towards neoliberal policies, with the free market increasingly being left to direct the shape of things, as discussed in earlier chapters. Since the early 1980s, therefore, there has been a shift away from a policy of containment of urbanisation to strategies which focus on increasing the overall productivity of urban areas (Harris, 1992; McGee, 1994).

This is most cogently reflected in the policy pronouncements of the major international aid agencies. Since the early 1990s, the World Bank as the major player has insisted on giving loans mainly to improve economic performance in the short term (Gould, 1992). Pugh (1995) describes the World Bank's post-1988 urban and housing policies as a 'quantum leap' from the 1970s and 1980s, and refers to this as the 'New Political Economy' (NPE), whilst others refer to the 'New Urban Management Programme' (Jones and Ward, 1994). The Bank places strong conditionalities on loans, relating them to desired economic, social, demographic and more recently environmental conditions as a part of Structural Adjustment Programmes. These have basically been designed to improve the macro-economic performance of countries (Harris, 1989).

The thinking behind this generic approach for the 1990s has been clearly stated in the World Bank Policy Paper, *Urban Policy and Economic Development: an Agenda for the 1990s* (World Bank, 1991; reproduced in Harris, 1992; see also Gould, 1992; Gaile, 1995; Rojas, 1995; Simon, 1995; Yeung, 1995). The policy document states that urban economic activities make up an increasing share of Gross Domestic Product in all nations. Thus, increasing the productivity of the urban economy is identified as the most salient policy variable and a more positive role is being ascribed to cities in the overall process of economic development. The United Nations Development Programme, for instance, comments that 'urbanization has been an essential part of most nations' development towards a stronger and more stable economy over the last few decades', and that 'the countries of the South that urbanized most rapidly in the last 10–20 years are generally also those with the most rapid economic growth' (UNDP, 1996: xxv). Following this line of argument, it is suggested that there is a major need to alleviate constraints to urban productivity.

The agenda notes how policies in the 1970s were primarily based on neighbourhood interventions, revolving around sites and services and slum upgrading schemes, together with municipal development and housing finance. Although based on principles of cost recovery and replicability, the policy paper states that these approaches had little or no city-wide impact. It is perceived now that the need is to focus on city-wide reforms relating to macro-economic policies, as well as on the broader objectives of economic development. In short, the World Bank has argued strongly that better macro-economic management policies are required.

Amongst the constraints to increased urban productivity identified are infrastructural deficiencies, what are seen as inappropriate urban regulatory policies governing construction, land and housing markets, the weakness of municipal institutions, and the inadequacy of financial services for urban development. These are recognised as the required focus if the urban economy is to prosper in the near future. It is maintained that just as at the micro-scale with housing upgrading and improvement, government needs to shift from a role as *provider* to one as *enabler*. It is envisaged that this will create a regulatory and financial framework within which private enterprise, households and the community can play an increasingly active role in meeting their own needs. Hence, the World Bank's urban activities are expected to grow, and the Bank is advocating a less regulated, more competitive free internal market system. Yet at the same time, the report notes the need to improve urban environmental conditions within developing countries.

In 1991 another major international organisation, the United Nations Development Programme (UNDP), produced a strategy paper under the title *Cities, People and Poverty: Urban Development Cooperation for the 1990s*. This also emphasised what it saw as the priority need to promote enabling strategies in seeking to alleviate poverty, provide the poor with shelter and service infrastructure, improve the urban environment and promote the private sector and Non-Governmental Organisations.

This is the overall context in which future urban development will occur, involving an acceptance of urban concentration, and more than that, a reasserted belief in the hegemony of urban-based productivity gains. Critics of the approach argue along the lines presented in earlier chapters, namely that such a commitment to market-led development and reliance on the private sector in conjunction with the over-ready acceptance of the value of urban-based projects will lead to increasing inequalities between social groups and between areas and regions. Likewise, unconstrained piecemeal developments are likely to see great pressures being placed on the environment. The argument that what has been happening amongst the Asian Tigers reflects a lack of unionisation, workers' rights and environmental regulations is often presented to the World Bank (see, for example, the extended discussion between World Bank staff and West Indian commentators on the economic scene in Wen and Sengupta, 1991). The social and environmental implications of such policies are very doubtful, as discussed throughout this text. The World Bank, together with the IMF, UNDP and the Inter-American Development Bank, seems to be increasingly aware of these arguments and is currently 'greening up' and acknowledging the need for ecological perspectives in development programmes (White, 1994).

The United Nations Centre for Human Settlements (Habitat) (UNCHS, 1966) noted recently that more than 600 million people in cities throughout the developing world are either homeless or reside in life- and health-threatening situations. The same source records that 'one of the greatest challenges for Habitat II is how governments can reduce this enormous health burden associated with poor quality housing' (UNCHS, 1996: xxviii), and that one of the great unresolved tasks is to explore 'how human settlement policies can help increase social equity, social integration, and social stability' (UNCHS, 1996: xxxi).

However, like it or not, in the majority of cases the state is being rolled back in this post-1989 non-Communist world, and there is an inexorable drift to the right in the urban policy arena (Potter and Unwin, 1995). More and more emphasis seems currently to be placed on Non-Governmental Organisations (Desai, 1995b). It remains to be seen how much these are able to achieve, in association with developments in Civil Society and new social movements, in alleviating the worst effects of poverty, inequality and environmental degradation, which seem bound to characterise the cities of the developing world over the coming decades.

References

Abrams, C., 1963, *Report to the Barbados Government and the Barbados Housing Authority on Land Tenure*, New York: Housing Policy and Home Finance.

Abrams, C., 1964, *Housing in the Modern World: Man's Struggle for Shelter in an Urbanizing World*, London: Faber and Faber.

Abu-Lughod, J., 1971, *Cairo: 1001 Years of the 'City Victorious'*, Princetown, NJ: Princetown University Press.

Action Aid, 1992, All in a life's work: the labour of children, *Common Cause*, 3(12), 1–26.

Adams, R.M., 1960, The origin of cities, *Scientific American* offprint, no. 606.

Adams, R.M., 1966, *The Evolution of Urban Society*, London: Weidenfeld & Nicolson.

Adams, W.M., 1990, *Green Development*, London: Routledge.

Afshar, H. (ed.), 1991, *Women, Development and Survival in the Third World*, Harlow: Longman.

Afshar, H. (ed.), 1996, *Women and Politics in the Third World*, London: Routledge.

Agarwal, B., 1992, The gender and environment debate: lessons for India, *Feminist Studies*, 18(1), 119–158.

Albornoz, O., 1993, *Education and Society in Latin America*, London: Macmillan.

Aldrich, B.C. and Sandhu, R.S. (eds), 1995, *Housing the Urban Poor: Policy and Practice in Developing Countries*, London and New Jersey: Zed.

Allen, J. and Hamnett, C. (eds), 1995, *A Shrinking World? Global Unevenness and Inequality*, Oxford: Oxford University Press in association with the Open University.

Allen, J. and Massey, D. (eds), 1995, *Geographical Worlds*, Oxford: Oxford University Press in association with the Open University.

Allen, T. and Thomas, A. (eds), 1992, *Poverty and Development in the 1990s*, Oxford and Milton Keynes: Oxford University Press and Open University Press.

Alonso, W., 1968, Urban and regional imbalances in economic development, *Economic Development and Cultural Change*, 17, 1–14.

Alonso, W., 1971, The economics of urban size, *Papers of the Regional Science Association*, 26, 67–83.

Alonso, W., 1980, Five bell shapes in development, *Papers of the Regional Science Association*, 45, 5–16.

Amato, P., 1970, Elitism and settlement patterns in the Latin American city, *Journal of the American Institute of Planners*, 36, 96–105.

Amato, P., 1971, A comparison of population densities, land values and socioeconomic class in four Latin American cities, *Land Economics*, 46, 447–455.

Amin, S., 1974, *Accumulation on a World Scale: A Critique of the Theory of Underdevelopment*, 2 volumes, New York: Monthly Review Press.

Amin, S., 1976, *Unequal Development*, New York: Monthly Review Press.

Amis, P., 1995, Making sense of urban poverty, *Environment and Urbanisation*, **7**, 145–157.

Amis, P. and Lloyd, P. (eds), 1990, *Housing Africa's Urban Poor*, Manchester: Manchester University Press.

Anderson, J., Brook, C. and Cochrane, A. (eds), 1996, *A Global World? Re-ordering Political Space*, Oxford: Oxford University Press in association with the Open University.

Andrews, F.M. and Phillips, G., 1970, The squatters of Lima: who are they and what do they want?, *Journal of Developing Areas*, **4**, 211–224.

Armstrong, W. and McGee, T.G., 1985, *Theatres of Accumulation: Studies in Asian and Latin American Urbanisation*, London: Methuen.

Auerbach, F., 1913, Das Gesetz der Bevölkreungskonzentration, *Petermann's Mitteilungen*, **59**, 74–76.

Augelli, J.P. and West, R.C., 1976, *Middle America: Its Land and Peoples*, Englewood Cliffs, NJ: Prentice-Hall.

Austin-Broos, D.J., 1995, Gay nights and Kingston Town: representations of Kingston, Jamaica, in Watson, S. and Gibson, K. (eds), *Postmodern Cities and Spaces*, Oxford: Blackwell, 149–164.

Auty, R.M., 1979, Worlds within the Third World, *Area*, **11**, 232–235.

Bairoch, P., 1975, *The Economic Development of the Third World since 1900*, London: Methuen.

Baker, J., 1995, Survival and accumulation strategies at the rural–urban interface in north-west Tanzania, *Environment and Urbanisation*, **7**, 117–132.

Ballara, M., 1992, *Women and Literacy*, London: Zed.

Barbier, E., 1989, *Economics, Natural Resource Scarcity and Development*, London: Earthscan.

Barnett, T. and Blaikie, P., 1992, *AIDS in Africa*, London: Belhaven.

Baross, P. and Linden, J.V., 1990, *The Transformation of Land Supply Systems in Third World Cities*, Aldershot: Avebury.

Beckford, G.L., 1972, *Persistent Poverty: Underdevelopment in Plantation Economies of the Third World*, New York: Oxford University Press.

Beneria, L. and Sen, G., 1981, Accumulation, reproduction and women's role in economic development, *Signs: Journal of Women in Culture and Society*, **7**(2), 279–298.

Benthall, J. and Corbridge, S.,1996, Urban–rural relations, demand politics and the 'new agrarianism' in northwest India: the Bharatiya Kisan Union, *Transactions of the Institute of British Geographers*, **21**, 27–48.

Bequele, A. and Boyden, J., 1988, Working children: current trends and policy responses, *International Labour Review*, **127**(2), 153–172.

Berghall, P.E., 1995, *Habitat II and the Urban Economy: A Review of Recent Developments and Literature*, Helsinki: United Nations University and the World Institute for Development Economics Research.

Berry, B.J.L., 1961, City size distributions and economic development, *Economic Development and Cultural Change*, **9**, 573–587.

Berry, B.J.L., 1973, Hierarchical diffusion: the basis of development filtering and spread in a system of growth centres, in Hansen, N.M. (ed.), *Growth Centres in Regional Economic Development*, New York: Free Press.

Berry, B.J.L. and Barnum, H.G., 1962, Aggregate relations and elemental components of central place systems, *Journal of Regional Science*, **4**, 35–68.

Betancur, J.J., 1987, Spontaneous settlement housing in Latin America: a critical examination, *Environment and Behaviour*, **19**, 286–310.

Bhatt, E. and the Self-Employed Women's Association (SEWA), 1989, Toward empowerment, *World Development*, **17**(7), 1059–1065.

Bird, J., 1977, *Centrality and Cities*, London: Routledge & Kegan Paul.

Blanes Jimenez, J., 1991, Cocaine, informality and the urban economy in La Paz, Bolivia, in Portes, A., Castells, M. and Benton, L., *The Informal Economy: Studies in Advanced and Less Developed Countries*, London: John Hopkins.

Borchert, J.R., 1967, American metropolitan evolution, *Geographical Review*, **57**, 301–323.

Boserup, E., 1970, *Women's Role in Economic Development*, London: Earthscan.

Boudeville, J.R., 1966, *Problems of Regional Economic Planning*, Edinburgh: Edinburgh University Press.

Boyden, J. and Holden, P., 1991, *Children of the Cities*, London: Zed.

Braidotti, R. *et al.* (eds), 1994, *Women, the Environment and Sustainable Development: Towards a Theoretical Synthesis*, London: Zed.

Brandt, W., 1980, *North–South: A Programme for Survival*, London: Pan.

Bray, M., Clarke, P.B. and Stephens, D., 1986, *Education and Society in Africa*, London: Arnold.

Breman, J., 1976, A dualistic labour system? A critique of the informal sector concept, I: The informal sector, *Economic and Political Weekly*, Nov. 27, 1870–1875.

Brierley, J., 1985, A review of development strategies and programmes of the People's Revolutionary Government in Grenada, 1979–83, *Geographical Journal*, **151**, 40–52.

Briggs, J., 1983, Rural development in Tanzania: end of an era?, *Geography*, **68**, 66–68.

Broad, R. and Landi, C.M., 1996, Whither the North–South gap?, *Third World Quarterly*, **17**, 7–17.

Bromley, R., 1978, The urban informal sector: critical perspectives, *World Development*, **6**, 1031–1198.

Bromley, R. and Gerry, C., 1979, *Casual Work and Poverty in Third World Cities*, Chichester: John Wiley.

Brookfield, H., 1975, *Interdependent Development*, London: Methuen.

Brunn, S.D. and Williams, J.F., 1983, *Cities of the World: World Regional Urban Development*, New York: Harper & Row.

Burgess, R., 1977, Self-help housing: a new imperialist strategy? A critique of the Turner school, *Antipode*, **9**, 50–59.

Burgess, R., 1978, Petty commodity housing or dweller control? A critique of John Turner's views on housing policy, *World Development*, **6**, 1105–1133.

Burgess, R., 1981, Ideology and urban residential theory in Latin America, in Herbert, D.T. and Johnston, R.J. (eds), *Geography and the Urban Environment: Progress in Research and Applications*, volume IV, Chichester: John Wiley, 57–114.

Burgess, R., 1982, Self-help housing advocacy: a curious form of radicalism: a critique of the work of John F.C. Turner, in Ward, P.M. (ed.), *Self-Help Housing: A Critique*, London: Mansell, 56–97.

Burgess, R., 1990, *The State and Self-Help Housing in Pereira, Colombia*, unpublished PhD thesis, University of London.

Burgess, R., 1992, Helping some to help themselves: Third World housing policies and development strategies, in Mathéy, K. (ed.), *Beyond Self-Help Housing*, London and New York: Mansell, 75–91.

Buttler, F.A., 1975, *Growth Pole Theory and Economic Development*, Farnborough: Saxon House.

Buvenic, M., 1983, Women's issues in Third World poverty: a policy analysis, in Buvenic, M., Lycette, M. and McGreevey, W.P. (eds), *Women and Poverty in the Third World*, Baltimore: Johns Hopkins University Press.

210

Cairncross, S., Hardoy, J.E. and Satterthwaite, D., 1990, *The Poor Die Young: Housing and Health in Third World Cities*, London: Earthscan.

Carby, H., 1983, White woman listen! Black feminism and the boundaries of sisterhood, in Centre for Contemporary Cultural Studies (ed.), *The Empire Strikes Back*, London: Hutchinson, 212–235.

Carol, H., 1964, Stages of technology and their impact upon the physical environment: a basic problem in cultural geography, *Canadian Geography*, **8**, 1–9.

Carter, H., 1977, Urban origins: a review, *Progress in Human Geography*, **1**, 12–32.

Carter, H., 1983, *An Introduction to Urban Historical Geography*, London: Arnold.

Castells, M., 1977, *The Urban Question: A Marxist Approach*, London: Arnold.

Castells, M., 1982, Squatters and politics in Latin America: a comparative analysis of urban social movements in Chile, Peru and Mexico, in Safa, H. (ed.), *Towards a Political Economy of Urbanisation in Third World Countries*, Delhi: Oxford University Press.

Castells, M. and Portes, A., 1991, World underneath: the origins, dynamics, and effects of the informal economy, in Portes, A., Castells, M. and Benton, L. (eds), *The Informal Economy: Studies in Advanced and Less Developed Countries*, London: John Hopkins.

CEPAL (Economic Commission for Latin America and the Caribbean), 1991, *La Equidad en el Panorama Social de America Latina durante los Anos Ochenta*, Santiago: CEPAL.

Cernea, M., 1993, The sociologist's approach to sustainable development, *Finance and Development*, **30**(4), 11–13.

Chambers, R., 1983, *Rural Development: Putting the Last First*, Harlow: Longman.

Chambers, R., 1988, Poverty in India: Concepts, research and reality, *IDS Discussion Paper*, no. 241, Institute of Development Studies, Sussex.

Chan, C., 1994, Responses of low-income communities to environmental challenges in Hong Kong, in Main, H. and Williams, S.W. (eds), *Environment and Housing in Third World Cities*, Chichester: Wiley.

Chandra, R., 1992, *Industrialization and Development in the Third World*, London and New York: Routledge.

Chang, S.-D., 1982, 'Modernization and urbanization problems in China', Occasional Paper 26, The Chinese University of Hong Kong.

Chant, S., 1987, Family structure and female labour in Querétaro, Mexico, in Momsen, J. and Townsend, J. (eds), *Geography of Gender in the Third World*, Hutchinson, London.

Chant, S., 1991, *Women and Survival in Mexican Cities: Perspectives on Gender, Labour Markets and Low-Income Households*, Manchester: Manchester University Press.

Chant, S., 1992, *Gender and Migration in Developing Countries*, London: Belhaven Press.

Chant, S., 1996a, Women's roles in recession and economic restructuring: Mexico and the Philippines, *Geoforum*, **23**(3), 297–327.

Chant, S., 1996b, *Gender, Urban Development and Housing*, United Nations Development Programme Publications Series for Habitat II, Volume 2, New York: UNDP.

Chant, S. and McIlwaine, C., 1995a, Gender and export-manufacturing in the Philippines: continuity or change in female employment? The case of Mactan Export processing zone, *Gender, Place and Culture*, **2**(2), 149–78.

Chant, S. and McIlwaine, C., 1995b, *Women of a Lesser Cost: Female Labour, Foreign Exchange and Philippine Development*, London: Pluto Press.

Cheema, G.S. and Rondinelli, D.A. (eds), 1983, *Decentralization and Development: Policy Implementation in Developing Countries*, London and New York: Sage.

Childe, V.G., 1950, The urban revolution, *Town Planning Review*, **21**, 3–17.

Childe, V.G., 1951, *Man Makes Himself*, New York: Merton.

Chokor, B.A., 1989, Motorway development and the conservation of traditional Third World cities, *Cities*, **6**(4), 317–324.

Christaller, W., 1933, *Die Zentralen Orte in Suddeutschland* (doctoral thesis). Translated by Baskin, C.W., 1966: *Central Places in Southern Germany*, Englewood Cliffs: Prentice Hall.

Chu-Sheng Lin, G., 1994, Changing theoretical perspectives on urbanisation in Asian developing countries, *Third World Planning Review*, **16**, 1–23.

Clark, G. and Dear, M., 1981, The state in capitalism and the capitalist state, in Dear, M. and Scott, A.J. (eds), *Urbanization and Urban Planning in Capitalist Society*, London and New York: Methuen.

Clarke, C.G., 1975, *Kingston, Jamaica: Urban Growth and Social Change, 1692–1962*, Berkeley, Los Angeles: University of California Press.

Clarke, C.G. and Ward, P., 1976, Stasis in make-shift housing: perspectives from Mexico and the Caribbean, *Actes du XLII Congrès International des Americanistes*, **10**, 351–358.

Clarke, J.A., 1982, The role of the state in regional development, chapter 6 in Flowerdew, R. (ed.), *Institutions and Geographical Patterns*, Beckenham: Croom Helm.

Clayton, A. and Potter, R.B., 1996, Industrial development and foreign direct investment in Barbados, *Geography*, **81**, 176–180.

Clegg, J., 1996, The development of multinational enterprises, chapter 7 in Daniels, P.W. and Lever, W.F. (eds), *The Global Economy in Transition*, Harlow: Addison Wesley Longman, 103–134.

Cleves Mosse, J., 1993, *Half the World, Half a Chance*, Oxford: Oxfam.

Cohen, R., 1987, *Helots: Migrants in the International Division of Labour*, Aldershot: Gower.

Connolly, P., 1985, The politics of the informal sector: a critique, in Mingione, E. and Redclift, N. (eds), *Beyond Employment*, London: Longman.

Conway, D., 1976, *Residential area change and residential relocation in Port of Spain, Trinidad*, unpublished PhD thesis, University of Texas at Austin.

Conway, D., 1981, Fact or opinion on uncontrolled peripheral settlement in Trinidad: or how different conclusions arise from the same data, *Ekistics*, **286**, 37–43.

Conway, D., 1982, Self-help housing, the commodity nature of housing and amelioration of the housing deficit: continuing the Turner–Burgess debate, *Antipode*, **14**, 40–46.

Conway, D., 1985, Changing perspectives on squatter settlements, intraurban mobility, and constraints on housing choice of the Third World urban poor, *Urban Geography*, **6**, 170–192.

Conyers, D., 1982, *An Introduction to Social Planning in the Third World*, Chichester: Wiley.

Cook, P. and Kirkpatrick, C., 1988, *Privatisation in Less Developed Countries*, Brighton: Harvester Wheatsheaf.

Cooke, P., 1990, Modern urban theory in question, *Transactions of the Institute of British Geographers*, New Series, **15**, 331–343.

Coote, B., 1995, *NAFTA: Poverty and Free Trade in Mexico*, Oxford: Oxfam.

Corbridge, S., 1991, Third world development, *Progress in Human Geography*, **15**(3), 311–321.

Corbridge, S., 1995, *Development Studies: A Reader*, London: Arnold.

Cornia, G.A., 1990, Investing in human resources: health, nutrition and development in the 1990s, in Griffin, K. and Knight, J. (eds), *Human Development and the International Development Strategy for the 1990s*, London: Macmillan.

Courtney, P.D. (ed.), 1994, *Geography and Development*, Melbourne: Longman Cheshire.

Crooke, P., 1983, Popular housing supports and the urban housing market, chapter 8 in Skinner, R.J. and Rodell, M.J. (eds), *People, Poverty and Shelter*, London: Methuen.

Cubitt, T., 1995, *Latin American Society*, 2nd edition, London: Longman.

Cumper, G.E., 1974, Dependence, development and the sociology of economic thought, *Social and Economic Studies*, **23**, 186–212.

Cuthbert, A., 1995, Under the volcano: postmodern space in Hong Kong, in Watson, S. and Gibson, K. (eds), *Postmodern Cities and Spaces*, Oxford: Blackwell, 138–148.

Daniels, P.W. and Lever, W.F. (eds), 1996, *The Global Economy in Transition*, London: Longman.

Dankleman, I. and Davidson, J., 1988, *Women and the Environment in the Third World*, London: Earthscan.

Dankleman, J. and Davidson, J., 1995, Human settlements: women's environment of poverty, in Kirby, J., O'Keefe, P. and Timberlake, L. (eds), *Sustainable Development: A Reader*, London: Earthscan.

Dann, G.M.S. and Potter, R.B., 1994, Tourism and postmodernity in a Caribbean setting, *Cahiers du Tourisme*, Series C, no. 185, 1–45.

Darwent, D.F., 1969, Growth poles and growth centres in regional planning – a review, *Environment and Planning*, **1**, 5–31.

Davies, R., 1979, The informal sector or subordinate mode of production? A model, in Bromley, R. and Gerry, C. (eds), *Casual Work and Poverty in Third World Cities*, Chichester: John Wiley.

Davis, K., 1965, The urbanization of the human population, *Scientific American*, **213**, 40–53.

Davis, K. (ed.), 1973, *Cities: Their Origin, Growth and Human Impact*, San Francisco: Freedman.

de Soto, H., 1989, *The Other Path: The Invisible Revolution in the Third World*, London: I.B. Taurus.

Dear, M. and Clark, G., 1978, The state and geographic process: a critical review, *Environment and Planning A*, **10**, 173–183.

Degg, M., 1994, Perspectives on urban vulnerability to earthquake hazard in the Third World, in Main, H. and Williams, S.W. (eds), *Environment and Housing in Third World Cities*, Chichester: Wiley.

Desai, V., 1995a, *Community Participation and Slum Housing: A Study of Bombay*, New Delhi, Thousand Oaks and London: Sage.

Desai, V., 1995b, *Filling the Gap: An Assessment of the Effectiveness of Urban NGOs*, Institute of Development Studies, University of Sussex.

Devas, N. and Rakodi, C. (eds), 1993, *Managing Fast Growing Cities: New Approaches to Urban Planning and Management in the Developing World*, Harlow: Longman.

Dicken, P., 1992, *Global Shift: The Internationalization of Economic Activity*, London: Paul Chapman.

Dickenson, J., 1994, Manufacturing industry in Latin America and the case of Brazil, in Courtney, P.P. (ed.), *Geography and Development*, Melbourne: Longman Cheshire, 165–191.

Dickenson, J.P. *et al.*, 1983, *A Geography of the Third World*, 1st edition, London: Routledge.

Dickenson, J.P. *et al.*, 1996, *A Geography of the Third World*, 2nd edition, London: Routledge.

Dixon, C., 1992, Human resources, in Dwyer, D. (ed.), *South East Asian Development: Geographical Perspectives*, Harlow: Longman.

Dixon, C.J. and Drakakis-Smith, D., 1995, The Pacific Asian region: myth or reality, *Geografiska Annaler*, **77B**(2), 75–91.

Dore, R., 1976, *The Diploma Disease: Education Qualification and Development*, London: Allen and Unwin.

Dos Santos, T., 1973, The crisis of development theory and the problem of dependence in Latin America, in Bernstein, H. (ed.), *Underdevelopment and Development in the Third World Today*, Harmondsworth: Penguin.

Douglas, I., 1983, *The Urban Environment*, London: Arnold.

Douglas, I., 1989, The environmental problems of cities, in Herbert, D.T. and Smith, D.M. (eds), *Social Problems and the City*, Oxford: Oxford University Press.

Douglass, M., 1992, The political economy of urban poverty and environmental management in Asia: access, empowerment and community based alternatives, *Environment and Urbanisation*, **4**(2), 9–32.

Doxiadis, C.A., 1967, Developments toward Ecumenopolis: the Great Lakes megalopolis, *Ekistics*, **22**, 14–31.

Doxiadis, C.A. and Papaiouannou, J.G., 1974, *Ecumenopolis: the Inevitable City of the Future*, New York: W.W. Norton.

Drakakis-Smith, D., 1981, *Urbanization, Housing and the Development Process*, London: Croom Helm.

Drakakis-Smith, D., 1987, *The Third World City*, London and New York: Methuen.

Drakakis-Smith, D., 1990, Food for thought or thought about food: urban food distribution systems in the Third World, in Potter, R.B. and Salau, A.T. (eds), *Cities and Development in the Third World*, London and New York: Mansell, 100–120.

Drakakis-Smith, D., 1992, Strategies for meeting basic food needs in Harare, in Baker, J. and Pedersen, P.O. (eds), *The Rural–Urban Interface in Africa*, Uppsala: Nordiska Afrikainstitutet.

Drakakis-Smith, D., 1993, Is there still a Third World?, *Choros 1993 1*, Department of Human and Economic Geography, Gothenburg University.

Drakakis-Smith, D., 1995, Third World cities: sustainable urban development I, *Urban Studies*, **32**(4–5), 659–677.

Drakakis-Smith, D., 1996a, Less developed economies and dependence, in Daniels, P.W. and Lever, W.F. (eds), *The Global Economy in Transition*, Harlow: Addison Wesley Longman, 215–238.

Drakakis-Smith, D., 1996b, Third World cities: sustainable development II – population, labour and poverty, *Urban Studies*, **33**, 673–701.

Drakakis-Smith, D., 1979, *High Society: Housing Provision in Metropolitan Hong Kong 1954–1979*, Centre of Asian Studies, University of Hong Kong.

Drakakis-Smith, D., 1991, Urban food distribution systems in Asia and Africa, *Geographical Journal*, **157**, 51–61.

Dunn, P., 1978, *Appropriate Technology*, Basingstoke: Macmillan.

Dwyer, D.J. (ed.), 1972, *The City as a Centre of Change in Asia*, Hong Kong: Hong Kong University Press.

Dwyer, D.J., 1975, *People and Housing in Third World Cities*, London: Longman.

Dwyer, D.J. (ed.), 1992, *South East Asian Development: Geographical Perspectives*, Harlow: Longman.

Dwyer, D.J., 1995, Urbanization, in Dwyer, D.J. (ed.), *South East Asian Development*, 2nd edition, Harlow: Longman.

Eckstein, S., 1988, *The Poverty of Revolution: The State and the Urban Poor in Mexico*, Princeton, NJ: Princeton University Press.

Eckstein, S., 1990, Urbanization revisited: inner-city slum of hope and squatter settlement of despair, *World Development*, **18**, 165–181.

Edwards, E.O. (ed.), 1974, *Employment in Developing Nations*, New York: Columbia University Press.

El-Shakhs, S., 1971, National factors in the development of Cairo, *Town Planning Review*, **42**, 235–249.

El-Shakhs, S., 1972, Development, primacy and systems of cities, *Journal of Developing Areas*, **7**, 11–36.

Elliott, J., 1994, *Sustainable Development*, London: Routledge.

Ellis, P. (ed.), 1986, *Women of the Caribbean*, London: Zed.

Elson, D., 1995, *Male Bias in the Development Process*, Manchester: Manchester University Press.

Evans, P., 1996, Government action, social capital and development: reviewing the evidence on synergy, *World Development*, **24**, 1119–1132.

Eyre, A., 1997, Self-help housing in Jamaica, in Potter, R.B. and Conway, D. (eds), *Self-Help Housing, the Poor and the State in the Caribbean*, Knoxville: University of Tennessee Press, 75–101.

Eyre, L.A., 1972, The shantytowns of Montego Bay, Jamaica, *Geographical Review*, **62**, 394–412.

Fagence, M., 1977, *Citizen Participation in Planning*, Oxford: Pergamon.

Fanstein, S. and Campbell, S. (eds), 1996, *Readings in Urban Theory*, Oxford: Blackwell.

Ferguson, B. and Maurer, C., 1996, Urban management for environmental quality in South America, *TWPR*, **18**, 117–154.

Ferguson, J., 1990, *Grenada: Revolution in Reverse*, London: Latin America Bureau.

Filani, M.O., 1981, Nigeria: the need to modify centre-down development planning, in Stöhr, W.B. and Taylor, D.R.F. (eds), *Development from Above or Below?*, Chichester: John Wiley, 283–304.

Filguera, C., 1983, To educate or not to educate: is that the question?, *CEPAL Review*, **21**.

Fiori, J. and Ramirez, R., 1992, Notes on the self-help housing critique, in Mathéy, K. (ed.), *Beyond Self-Help Housing*, London and New York: Mansell, 23–31.

Firman, T., 1996, Urban development in Bandung Metropolitan Region: a transformation to a Desa-kota region, *Third World Planning Review*, **18**, 1–22.

Forbes, D. and Thrift, N., 1987, *The Socialist Third World: Urban Development and Territorial Planning*, Oxford: Blackwell.

Foster, P., 1992, *The World Food Problem*, Boulder, Co.: Lynne Rienner.

Frank, A.G., 1966, The development of underdevelopment, *Monthly Review*, September 1996, 17–30.

Frank, A.G., 1967, *Capitalism and Underdevelopment in Latin America*, New York: Monthly Review Press.

Frank, A.G., 1980, *Crisis in the World Economy*, New York: Holmes & Meier.

Friedmann, J., 1966, *Regional Development Policy: A Case Study of Venezuela*, Massachusetts: MIT Press.

Friedmann, J., 1986, The world city hypothesis, *Development and Change*, **17**, 69–83.

Friedmann, J., 1995, Where we stand: a decade of world city research, chapter 2 in Knox, P.L. and Taylor, P.J. (eds), *World Cities in a World-System*, Cambridge: Cambridge University Press.

Friedmann, J. and Douglas, M., 1975, Agropolitan development: toward a new strategy for regional planning in Asia, in Lo, F.C. and Salih, K. (eds), *Growth Pole Strategy and Regional Development Policy: Asian Experience and Alternative Approaches*, Oxford: Pergamon.

Friedmann, J. and Weaver, C., 1979, *Territory and Function: the Evolution of Regional Planning*, London: Arnold.

Friedmann, J. and Wolff, G., 1982, World city formation: an agenda for research and action, *International Journal of Urban and Regional Research*, **6**, 309–343.

Friedmann, J. and Wulff, R., 1976, *The Urban Transition: Comparative Studies of Newly Industrializing Societies*, London: Arnold.

Frobel, F., Heinrichs, J. and Kreye, O., 1980, *The New International Division of Labour*, Cambridge: Cambridge University Press.

Furedy, C. and Alamgir, M., 1992, Street pickers in Calcutta slums, *Environment and Urbanisation*, **4**(2), 54–58.

Fyfe, A., 1994, *Child Labour*, Cambridge: Polity.

Gaile, G.L., 1995, Commentary: an underview of Cedric Pugh, *Cities*, **12**, 405–407.

George, S., 1988, *A Fate Worse Than Debt*, Harmondsworth: Penguin.

George, S. and Sabelli, F., 1994, *Faith and Credit: The World Bank's Secular Empire*, Harmondsworth: Penguin.

Ghazi, P., 1993, America's deadly border, *The Observer Magazine*, Sunday 12 December, 16–20.

Gibbon, P., 1996, Structural adjustment and structural change in sub-Saharan Africa: some provisional conclusions, *Development and Change*, **27**, 751–784.

Gilbert, A., 1983, The tenants of self-help housing: choice and constraint in the housing markets of less developed countries, *Development and Change*, **14**, 449–477.

Gilbert, A., 1990, *Latin America*, London: Routledge.

Gilbert, A., 1992, Third World cities: housing, infrastructure and servicing, *Urban Studies*, **29**, 435–460.

Gilbert, A., 1993a, Third World cities: the changing national settlement system, *Urban Studies*, **30**, 721–740.

Gilbert, A., 1993b, *In Search of a Home: Rental and Shared Housing in Caracas, Santiago and Mexico City*, London: UCL Press.

Gilbert, A., 1994, Third World cities: poverty, employment, gender roles and the environment during a time of restructuring, *Urban Studies*, **31**, 605–633.

Gilbert, A., 1996, *The Mega-City in Latin America*, Tokyo, New York, Paris: United Nations University Press.

Gilbert, A. and Healey, P., 1985, *The Political Economy of Land: Urban Development in an Oil Economy*, Aldershot: Gower.

Gilbert, A. and Varley, A., 1991, *Landlord and Tenant: Housing the Poor in Urban Mexico*, London: Routledge.

Gilbert, A. and Ward, P., 1982, Residential movement among the poor: the constraints on houisng choice in Latin American cities, *Transactions of the Institute of British Geographers*, **7**(2), 129–149.

Gilbert, A.G., 1976, The arguments for very large cities reconsidered, *Urban Studies*, **13**, 27–34.

Gilbert, A.G., 1977, The argument for very large cities reconsidered: a reply, *Urban Studies*, **14**, 225–227.

Gilbert, A.G., 1981, Pirates and invaders: land acquisition in urban Colombia and Venezuela, *World Development*, **9**, 657–678.

Gilbert, A.G. and Goodman, D.E., 1976, Regional income disparities and economic development, in Gilbert, A.G. (ed.), *Development Planning and Spatial Structure*, Chichester: Wiley.

Gilbert, A.G. and Gugler, J., 1992, *Cities, Poverty and Development: Urbanization in the Third World*, 2nd edition, Oxford: Oxford University Press.

Ginsburg, N., Koppell, B. and McGee, T.G. (eds), 1991, *The Extended Metropolis: Settlement Transition in Asia*, Honolulu: University of Hawaii Press.

Girvan, N., 1973, The development of dependency economics in the Caribbean and Latin America: review and comparison, *Social and Economic Studies*, **22**, 1–33.

Goldschmidt-Clermont, L., 1990, Economic measurement of non-market household activities. Is it useful and feasible?, *International Labour Review*, **129**(3), 279–302.

Goodman, D. and Redclift, M. (eds), 1991, *Environment and Development in Latin America: The Politics of Sustainability*, Manchester: Manchester University Press.

Gordon, D., 1989, The global economy: new edifice or crumbling foundations?, *New Left Review*, **168**, 24–65.

Gottmann, J., 1957, Megalopolis or urbanization of the north-eastern seaboard, *Economic Geography*, **33**, 189–200.

Gottmann, J., 1978, Megalopolitan systems around the world, in Bourne, L.S. and Simmons, J.W. (eds), *Systems of Cities*, Oxford: Oxford University Press, 53–60.

Gottmann, J., 1983, Third World cities in perspective, *Area*, **15**, 311–313.

Gould, P., 1969, The structure of space preferences in Tanzania, *Area*, **1**, 29–35.

Gould, P.R., 1970, Tanzania, 1920–63: the spatial impress of the modernisation process, *World Politics*, **22**, 149–170.

Gould, W.T.S., 1986, Population analysis for the planning of primary schools in the Third World, in Gould, W.T.S. and Lawton, R. (eds), *Planning for Population Change*, London: Croom Helm.

Gould, W.T.S., 1992, Urban development and the World Bank, *Third World Planning Review*, **14**, iii–vi.

Gould, W.T.S., 1993, *People and Education in the Third World*, Harlow: Longman.

Graham-Brown, S., 1991, *Education in the Developing World: Conflict and Crisis*, Harlow: Longman.

Grant, M., 1995, Movement patterns and medium sized city: tenants on the move in Gweru, Zimbabwe, *Habitat International*, **19**, 357–370.

Green, D., 1995, *Silent Revolution: The Rise of Market Economics in Latin America*, London: Cassell.

Griffin, E. and Ford, L., 1980, A model of Latin American city structure, *Geographical Review*, **70**, 397–422.

Guardian, 1993, Toy factory folds like house of cards in Bangkok inferno, *Guardian*, Wednesday 12 May.

Guardian, 1994, Out of sight and out of mind, *Guardian*, Monday 14 March, 12–13.

Gugler, J., 1980, A minimum of urbanism and a maximum of ruralism: the Cuban experience, *International Journal of Urban and Regional Research*, **4**, 516–535.

Gwynne, R.N., 1978, City size and retail prices in less-developed countries: an insight into primacy, *Area*, **10**, 136–140.

Hagerstrand, T., 1953, *Innovationsforlopet ur Korologisk Synpunkt*, Lund, Sweden.

Hall, D.R., 1981a, External relations and current development patterns in Cuba, *Geography*, **66**, 237–240.

Hall, D.R., 1981b, Town and country planning in Cuba, *Town and Country Planning*, **50**, 81–83.

Hall, D.R., 1989, Cuba, in Potter, R.B. (ed.), *Urbanization, Planning and Development in the Caribbean*, London and New York: Mansell.

Hall, M., 1992, Sex tourism in South-East Asia, in Harrison, D. (ed.), *Tourism and the Less Developed Countries*, London: Belhaven.

Hall, P., 1966, *The World Cities*, London: Weidenfeld & Nicolson.

Hamberg, J., 1990, Cuba, in Mathéy, K. (ed.), *Housing Policies in the Socialist Third World*, London: Mansell, 35–70.

Hamdi, N., 1991, *Housing Without Houses: Participation, Flexibility, Enablement*, New York: Van Nostrand Reinhold.

Hamnett, C., 1995, Controlling space: global cities, in Allen, J. and Hamnett, C. (eds), *A Shrinking World? Global Unevenness and Inequality*, Oxford: Oxford University Press in association with the Open University.

Hansen, N.M., 1967, Development pole theory in a regional context, *Kyklos*, **20**, 709–725.

Hansen, N.M. (ed.), 1972, *Growth Centres in Regional Economic Development*, New York: Free Press.

Hansen, N.M., 1981, Development from above: the centre-down development paradigm, in Stöhr, W.B. and Taylor, D.R.F. (eds), *Development from Above or Below?*, Chichester: John Wiley.

Hanson, S. and Pratt, G., 1995, *Gender, Work and Space*, London: Routledge.

Harcourt, W., 1994a, The globalisation of the economy: an international gender perspective, *Focus on Gender*, **2**(3), 6–14.

Harcourt, W. (ed.), 1994b, *Feminist Perspectives on Sustainable Development*, London: Zed.

Hardoy, J.E. and Satterthwaite, D. (eds), 1986, *Small and Intermediate Urban Centres: Their Role in National and Regional Development in the Third World*, London: Hodder & Stoughton.

Hardoy, J.E. and Satterthwaite, D., 1989, *Squatter Citizen: Life of the Urban Third World*, London: Earthscan.

Hardoy, J.E. and Satterthwaite, D., 1990, The future city, in Cairncross, S., Hardoy, J.E. and Satterthwaite, D. (eds), *The Poor Die Young*, London: Earthscan.

Hardoy, J.E. and Satterthwaite, D., 1995, Urban growth as a problem, in Kirby, J., O'Keefe, P. and Timberlake, L. (eds), *Sustainable Development: A Reader*, London: Earthscan.

Hardoy, J.E., Mitlin, D. and Satterthwaite, D., 1992, *Environmental Problems in Third World Cities*, London: Earthscan.

Hardy, D. and Ward, C., 1984, *Arcadia for All: The Legacy of a Makeshift Landscape*, London: Mansell.

Harms, H., 1982, Historical perspectives on the practice and purpose of self-help housing, in Ward, P.M. (ed.), *Self-Help Housing: A Critique*, London: Mansell.

Harms, H., 1992, Self-help housing in developed and Third World countries, in Mathéy, K. (ed.), *Beyond Self-Help Housing*, London and New York: Mansell, 33–52.

Harris, N., 1989, Aid and urbanization, *Cities*, **6**, 174–185.

Harris, N., 1992, *Cities in the 1990s: the Challenge for Developing Countries*, London: UCL Press.

Hart, K., 1973, Informal income opportunities and urban employment in Ghana, *Journal of Modern African Studies*, **11**, 61–89.

Harvey, D., 1973, *Social Justice and the City*, London: Arnold.

Harvey, D., 1989, *The Condition of Postmodernity*, Oxford: Blackwell.

Herz, B. and Measham, A., 1987, *The Safe Motherhood Initiative: Proposals for Action*, Washington, DC: World Bank.

Hettne, B., 1994, *Development Theory and the Three Worlds*, 2nd edition, Harlow: Longman.

Hewitt, T., Johnson, H. and Wield, D. (eds), 1992, *Industrialization and Development*, Oxford: Oxford University Press in association with the Open University.

Hilling, D., 1996, *Transport and Developing Countries*, London and New York: Routledge.

Hirschman, A.O., 1958, *The Strategy of Economic Development*, New Haven, CT: Yale University Press.

Hirst, M., 1978, Recent villagization in Tanzania, *Geography*, **63**, 122–125.

Hoch, I., 1972, Income and city size, *Urban Studies*, **9**, 299–328.

Hollier, G.P., 1988, Regional development, in Pacione, M. (ed.), *The Geography of the Third World: Progress and Prospects*, London and New York: Routledge, 232–270.

Holmberg, J. and Sandbrook, R., 1992, Sustainable development: what is to be done?, in Holmberg, J. (ed.), *Policies for a Small Planet*, London: Earthscan.

Hope, K.R. and Ruefli, T., 1981, Rural–urban migration and the development process: a Caribbean case study, *Labour and Society*, **6**, 145–146.

Horvath, R.J., 1969, In search of a theory of urbanisation: notes on the colonial city, *East Lakes Geographer*, **5**, 69–82.

Horvath, R.J., 1972, A definition of colonialism, *Current Anthropology*, **13**, 45–57.

Hoselitz, B.F., 1955, Generative and parasitic cities, *Economic Development and Cultural Change*, **3**, 278–294.

Hoyle, B.S., 1979, African socialism and urban development: the relocation of the Tanzanian capital, *Tijdschrift voor Economische en Sociale Geografie*, **70**, 207–216.

Hoyle, B.S., 1983, *Seaports and Development: the Experience of Kenya and Tanzania*, London: Gordon & Breach.

Hoyt, H., 1939, *The Structure and Growth of Residential Neighbourhoods in American Cities*, Washington, DC: Federal Housing Administration.

Hudson, B., 1989, The Commonwealth Eastern Caribbean, in Potter, R.B. (ed.), *Urbanization, Planning and Development in the Caribbean*, London and New York: Mansell, 181–211.

Hudson, B., 1991, Physical planning in the Grenada Revolution: achievement and legacy, *Third World Planning Review*, **13**, 179–190.

Hudson, J.C., 1969, Diffusion in a central place system, *Geographical Analysis*, **1**, 45–58.

Hussman, R., 1990, *Surveys of Economically Active Population, Employment, Unemployment and Underemployment: An ILO Manual on Concepts and Methods*, Geneva: International Labour Office Publications.

International Labour Office, 1970, *Towards Full Employment: A Programme for Columbia*, Geneva: International Labour Office Publications.

International Labour Office, 1972, *Employment, Incomes and Equality: A Strategy for Increasing Productive Employment in Kenya*, Geneva: International Labour Office Publications.

International Labour Office, 1988, *World Employment Review*, Geneva: International Labour Office.

International Labour Office, 1995, *World Employment 1995*, Geneva: International Labour Office.

International Labour Organization, 1976a, *Declaration of Principles and Programme of Action for a Basic Needs Strategy of Development*, Geneva: International Labour Office.

International Labour Organization, 1976b, *Employment, Growth and Basic Needs: A One-World Problem*, Geneva: International Labour Office.

Jackson, C., 1993, Environmentalisms and gender interests in the Third World, *Development and Change*, **24**, 649–677.

Jackson, P., 1987, The idea of 'race' and the geography of racism, in Jackson, P. (ed.), *Race and Racism*, London: Allen and Unwin, 3–21.

Jacobs, J., 1969, *The Economy of Cities*, New York: Random House.

Jaising, I. and Sathyamala, C., 1995, Legal rights . . . and wrongs: internationalising Bhopal, in Kirby, J., O'Keefe, P. and Timberlake, L. (eds), *Sustainable Development: A Reader*, London: Earthscan.

James, C.L.R., 1963, *Beyond a Boundary*, New York: Pantheon.

Jameson, F., 1984, Postmodernism or the cultural logic of late capitalism, *New Left Review*, **146**, 53–92.

Jefferson, M., 1939, The law of the primate city, *Geographical Review*, **20**, 226–232.

Jenkins, R., 1987, *Transnational Corporations and Uneven Development*, London: Methuen.

Jenkins, R., 1992, Industrialization and the global economy, in Hewitt, T., Johnson, H. and Wield, D. (eds), *Industrialization and Development*, Oxford: Oxford University Press in association with the Open University, 13–40.

Jimenez-Diaz, V., 1994, The incidence and causes of slope failures in the barrios of Caracas, Venezuela, in Main, H. and Williams, S.W. (eds), *Environment and Housing in Third World Cities*, Chichester: Wiley.

Johnson, B.L.C., 1983, *India: Resources and Development*, 2nd edition, London: Heinemann.

Johnson, E.A.J., 1965, *Market Towns and Spatial Development in India*, New Delhi: Oxford University Press.

Johnson, E.A.J., 1970, *The Organization of Space in Developing Countries*, New York: Harvard University Press.

Johnston, R.J., 1971, On the progression from primacy to rank-size in an urban system: the deviant case of New Zealand, *Area*, **3**, 180–184.

Johnston, R.J., 1977, Regarding urban origins, urbanization and urban patterns, *Geography*, **62**, 1–8.

Johnston, R.J., 1980, *City and Society: An Outline for Urban Geography*, Harmondsworth: Penguin.

Johnston, R.J. and Taylor, P.J., 1986, *A World in Crisis?*, Oxford: Basil Blackwell.

Johnston, R.J., Taylor, P.J. and Watts, M.J., 1996, *Geographies of Global Change*, Oxford: Blackwell.

Jones, E. and Eyles, J., 1977, *An Introduction to Social Geography*, Oxford: Oxford University Press.

Jones, G. and Ward, P.M. (eds), 1994, *Methodology for Land and Housing Market Analysis*, London: UCL Press.

Jones, G.A., 1996, The difference between truth and adequacy: (re)joining Baken, van der Linden and Malpezzi, *Third World Planning Review*, **18**, 243–256.

Jones, T., 1988, *Corporate Killing: Bhopals will Happen*, London: Free Association Books.

Kaarsholm, P. (ed.), 1995, *From Post-Traditional to Post-Modern? Interpreting the Meaning of Modernity in Third World Urban Societies*, Occasional Paper no. 14/1995, International Development Studies, Roskilde University.

Kabeer, N., 1994, *Reversed Realities: Gender Hierarchies and Development Thought*, London: Verso.

Kasada, J.D. and Parnell, A.M. (eds), 1993, *Third World Cities*, London: Sage.

Kellett, P., 1995, *Constructing home: Production and consumption of popular housing in northern Colombia*, unpublished PhD thesis, University of Newcastle upon Tyne.

Keung, J.K., 1985, Government intervention and housing policy in Hong Kong, *Third World Planning Review*, **7**, 23–44.

King, A.D., 1976, *Colonial Urban Development: Culture, Social Power and Environment*, London: Routledge and Kegan Paul.

King, A.D., 1991, *Culture, Globalization and the World System*, Basingstoke: Macmillan.

Kirby, J., O'Keefe, P. and Timberlake, L., 1995, *Sustainable Development: A Reader*, London: Earthscan.

Kirkby, R.J.R., 1985, *Urbanisation in China: Town and Country in a Developing Economy, 1949–2000*, London: Croom Helm.

Kirton, C.D., 1988, Public policy and private capital in the transition to socialism: Grenada 1979–93, *Social and Economic Studies*, **37**, 125–150.

Knox, P., 1996, World cities and the organisation of global space, in Jonhston, R.J., Taylor, P. and Watts, M. (eds), *Geographies of Global Change: Remapping the World in the Late Twentieth Century*, Oxford: Blackwell.

Knox, P.L. and Taylor, P.J. (eds), 1995, *World Cities in a World-System*, Cambridge: Cambridge University Press.

Koningsburger, O.H., 1983, The role of the planner in a poor (and in a not quite so poor) country, *Habitat International*, **7**, 49–55.

Korten, N., 1996, Civic engagement in creating future cities, *Environment and Urbanisation*, **8**, 35–49.

Krausse, G.H., 1978, Intra-urban variation in Kampung settlements of Jakarta: a structural analysis, *Journal of Tropical Geography*, **46**, 11–46.

Kuklinski, A. (ed.), 1972, *Growth Poles and Growth Centres in Regional Planning*, Mouton.

Kumar, S., 1996, Landlordism in Third World urban low income settlements: a case for further research, *Urban Studies*, **33**, 753–782.

Kuznets, S., 1955, Economic growth and income inequality, *American Economic Review*, **45**, 1–28.

La Guerre, J. (ed.), 1993, *Structural Adjustment: Public Policy and Administration in the Caribbean*, St Augustine, Trinidad and Tobago: School of Continuing Studies, University of the West Indies.

Laguerre, M.S., 1990, *Urban Poverty in the Caribbean*, Basingstoke: Macmilllan.

Lampard, E.E., 1955, The history of cities in the economically advanced areas, *Economic Development and Cultural Change*, **3**, 81–102.

Langton, J., 1975, Residential patterns in pre-industrial cities: some case studies of the seventeenth century, *Transactions of the Institute of British Geographers*, **65**, 1–27.

Laquian, A.A., 1983, Sites, services and shelter – an evaluation, *Habitat International*, **7**, 211–225.

Lasuén, J.R., 1969, On growth poles, *Urban Studies*, **6**, 137–161.

Lasuén, J.R., 1973, Urbanisation and development – the temporal interaction between geographical and sectoral clusters, *Urban Studies*, **10**, 163–188.

Lawson, V., 1992, Industrial subcontracting and employment forms in Latin America: a framework for contextual analysis, *Progress in Human Geography*, **16**, 1–23.

Le Franc, E., 1989, Petty trading and labour mobility: Higglers in the Kingston metropolitan area, in Hart, K. (ed.), *Women and the Sexual Division of Labour in the Caribbean*, Mona, Jamaica: Consortium Graduate Scool of Social Sciences, University of the West Indies.

Lee, W., 1991, Prostitution and tourism in South-East Asia, in Redclift, N. and Sinclair, T.M. (eds), *Working Women: International Perspectives on Labour and Gender Ideology*, London: Routledge.

Lee-Smith, D., 1990, Squatter landlords in Nairobi: a case study of Korogocho, in Amis, P. and Lloyd, P. (eds), *Housing Africa's Urban Poor*, Manchester: Manchester University Press, 175–188.

Lehmann, D., 1982, Agrarian structure, migration and the state in Cuba, in Peck, P. and Standing, G. (eds), *State Policies and Migration: Studies in Latin America and the Caribbean*, London: Croom Helm, 321–390.

Leinbach, T.R., 1972, The spread of modernization in Malaya, 1895–1969, *Tijdschrift voor Economische en Sociale Geografie*, **63**, 262–277.

Leitmann, J., Bartone, C. and Bernstein, J., 1992, Environmental management and urban development: issues and options for Third World cities, *Environment and Urbanisation*, **4**(2), 131–140.

Lewis, O., 1959, *Five Families: Mexican Case Studies in the Culture of Poverty*, New York: Basic Books.

Lewis, O., 1966, The culture of poverty, *Scientific American*, **215**, 19–25.

Lewis, W.A., 1950, The industrialisation of the British West Indies, *Caribbean Economic Review*, **2**, 1–61.

Lewis, W.A., 1954, Economic development with unlimited supplies of labour, *Manchester School of Economic and Social Studies*, **22**(2), 139–191.

Lewis, W.A., 1955, *The Theory of Economic Growth*, London: George Allen & Unwin.

Linsky, A.S., 1965, Some generalisations concerning primate cities, *Annals of the Association of American Geographers*, **55**, 506–513.

Lipton, M., 1977, *Why Poor People Stay Poor: Urban Bias in World Development*, London: Temple Smith.

Lipton, M., 1980, Migration from rural areas of poor countries: the impact on rural productivity and income distribution, *World Development*, **8**, 1–24.

Lipton, M., 1982, Why poor people stay poor, in Harriss, J. (ed.), *Rural Development: Theories of Peasant Economy and Agrarian Change*, London: Hutchinson.

Lloyd, P., 1979, *Slums of Hope? Shanty Towns of the Third World*, Harmondsworth: Penguin.

Lloyd-Evans, S., 1994, *Ethnicity, gender and the informal sector in Trinidad: with particular reference to petty-commodity trading*, unpublished PhD thesis, University of London.

Lloyd-Evans, S., 1995, Prospects for sustainable job creation: gender, ethnicity and the informal sector in Trinidad, University of Reading *Geographical Papers Series B*, **31**, 1–22.

Lloyd-Evans, S. and Potter, R.B., 1992, The informal sector of the economy in the Commonwealth Caribbean: an overview, *Bulletin of Eastern Caribbean Affairs*, **17**, 26–40.

Lloyd-Evans, S. and Potter, R.B., 1993, Government response to informal sector retailing trading: the People's Mall, Port of Spain, Trinidad, *Geography*, **78**(3), 315–318.

Lloyd-Evans, S. and Potter, R.B., 1996, Environmental impacts of urban development and the urban informal sector in the Caribbean, in Eden, M. and Parry, J.T. (eds), *Land Degradation in the Tropics*, London: Pinter.

Lo, F. and Salih, K., 1981, Growth poles, agropolitan development and polarization reversal: the debate and search for alternatives, in Stöhr, W.B. and Taylor, D.R.F. (eds), *Development From Above or Below?*, Chichester: Wiley.

Lohmann, L., 1993, Green orientalism, *The Ecologist*, **23**(6), 202–204.

Lomnitz, L., 1977, *Networks and Marginality: Life in a Mexican Shanty Town*, Academic Press.

Lösch, A., 1940, *Die räumliche Ordnung der Wirtschaft*, Jena, translated by Woglom, W.H. and Stolper, W.F., 1954, *The Economics of Location*, Yale University Press.

Lowder, S., 1986, *Inside Third World Cities*, Beckenham: Croom Helm.

Lowder, S., 1991, The context of urban planning in secondary cities, *Cities*, **8**, 54–65.

Lowenthal, D., 1960, *West Indian Societies*, Oxford: Oxford University Press.

Lundqvist, J., 1981, Tanzania: socialist ideology, bureaucratic reality and development from below, chapter 13 in Stöhr, W.B. and Taylor, D.R.F. (eds), *Development From Above or Below?*, Chichester: Wiley, 329–349.

Ma, L.J.C., 1976, Anti-urbanism in China, *Proceedings of the Association of American Geographers*, **8**, 114–118.

Ma, L.J.C. and Noble, A.G., 1986, Chinese cities: a research agenda, *Urban Geography*, **7**, 279–290.

Mabogunje, A.L., 1980, *The Development Process: A Spatial Perspective*, London: Hutchinson.

Macdonald, M., 1994, *Gender Planning in Development Agencies*, Oxford: Oxfam.

MacGregor, M.T.G. and Valverde, V.C., 1975, Evolution of the urban population in the arid zones of Mexico, 1900–1970, *Geographical Review*, **65**, 214–228.

MacLeod, S. and McGee, T., 1990, The last frontier: the emergence of the industrial palate in Hong Kong, chapter 11 in Drakakis-Smith, D. (ed.), *Economic Growth and Urbanization in Developing Areas*, London and New York: Routledge.

Main, H., 1994, Introduction, in Main, H. and Williams, S.W. (eds), *Environment and Housing in Third World Cities*, Chichester: Wiley.

Main, H. and Williams, S.W. (eds), 1994, *Environment and Housing in Third World Cities*, Chichester: Wiley.

Main, H.A.C., 1990, Housing problems and squatter solutions in metropolitan Kano, in Potter, R.B. and Salau, A.T. (eds), *Cities and Development*, London and New York: Mansell.

Mangin, W., 1967, Latin American squatter settlements: a problem and a solution, *Latin American Research Reviews*, **2**, 65–98.

Mannion, A.M., 1992, Sustainable development and biotechnology, *Environmental Conservation*, **19**(4), 298–305.

Marcuse, P., 1992, Why conventional self-help projects won't work, in Mathéy, K. (ed.), *Beyond Self-Help Housing*, London and New York: Mansell, 15–21.

Martin, R.J., 1983, Upgrading, in Skinner, R.J. and Rodell, M. (eds), *People, Poverty and Shelter: Problems of Self-Help Housing in the Third World*, London and New York: Mansell.

Masselos, J., 1995, Postmodern Bombay: fractured discourse, in Watson, S. and Gibson, K. (eds), *Postmodern Cities and Spaces*, Oxford: Blackwell, 200–215.

Massey, D., 1991, A global sense of place, *Marxism Today*, June 1991, 24–29.

Massey, D. and Jess, P., 1995, *A Place in the World? Places, Cultures and Globalization*, Oxford: Oxford University Press in association with the Open University.

Massiah, J., 1989, Women's lives and livelihoods: a view from the Commonwealth Caribbean, *World Development*, **17**(7), 965–977.

Massiah, J. (ed.), 1993, *Women in Developing Economies: Making Visible the Invisible*, Oxford: Berg/UNESCO.

Mathéy, K. (ed.), 1992a, *Beyond Self-Help Housing*, London and New York: Mansell.

Mathéy, K., 1992b, Self-help housing policies and practices in Cuba, in Mathéy, K. (ed.), *Beyond Self-Help Housing*, London and New York: Mansell, 181–216.

Mathéy, K., 1992c, Positions on self-help housing: bibliographic notes, in Mathéy, K. (ed.), *Beyond Self-Help Housing*, London and New York: Mansell, 379–396.

Mathéy, K., 1997, Self-help housing strategies in Cuba: an alternative to conventional wisdom?, in Potter, R.B. and Conway, D. (eds), *Self-Help Housing, the Poor and the State in the Caribbean*, Knoxville: University of Tennessee Press, 164–187.

Maynard, M., 1994, 'Race', gender and the concept of 'difference' in feminist thought, in Afshar, H. and Maynard, M. (eds), *The Dynamics of 'Race' and Gender: Some Feminist Interventions*, London: Taylor and Francis, 9–21.

McAfee, K., 1991, *Storm Signals: Structural Adjustment and Development Alternatives in the Caribbean*, London: Zed.

McAuslan, P., 1985, *Urban Land and Shelter for the Poor*, London and Washington: International Institute for Environment and Development.

McCormack, C., 1980, Nature, culture and gender: a critique, in McCormack, C. and Strathern, M. (eds), *Nature, Culture and Gender*, Cambridge: Cambridge University Press.

McElroy, J.J. and Albuquerque, K., 1986, The tourism demonstration effect on the Caribbean, *Journal of Travel Research*, **25**, 31–34.

McGee, T., 1979a, Conservation and dissolution in the Third World city: the 'shanty town' as an element of conservation, *Development and Change*, **10**, 1–22.

McGee, T., 1994, The future of urbanisation in developing countries: the case of Indonesia, *Third World Planning Review*, **16**, iii-xii.

McGee, T.G., 1967, *The South East Asian City*, London: Bell.

McGee, T.G., 1971, *The Urbanization Process in the Third World: Exploration in Search of a Theory*, London: Bell.

McGee, T.G., 1978, An invitation to the ball. Dress 'formal' or 'informal'?, in Rimmer, P., Drakakis-Smith, D. and McGee, T.G. (eds), *Studies in Food, Shelter and Transport in the Third World*, Canberra: Australian National University.

McGee, T.G., 1979, The poverty syndrome: making out in the southeast Asian city, in Bromley, R. and Gerry, C. (eds), *Casual Work and Poverty in Third World Cities*, Chichester: John Wiley.

McGee, T.G., 1989, Urbanisasi or Kotadesasi? Evolving patterns of urbanisation in Asia, in Costa, F.J. (ed.), *Urbanization in Asia: Spatial Dimensions and Policy Issues*, Honolulu: University of Hawaii Press, 93–110.

McGee, T.G., 1991a, Eurocentrism in geography: the case of Asian urbanization, *Canadian Geographer*, **35**, 332–344.

McGee, T.G., 1991b, The emergence of Desakota regions in Asia: expanding a hypothesis, in Ginsburg, N., Koppell, B. and McGee, T.G. (eds), *The Extended Metropolis: Settlement Transition in Asia*, Honolulu: University of Hawaii Press, 3–25.

McGee, T.G., 1995, Eurocentrism and geography: reflections on Asian urbanisation, in Crush, J. (ed.), *Power of Development*, London: Routledge.

McGee, T.G. and Greenberg, C., 1992, The emergence of extended metropolitan regions in ASEAN, *ASEAN Economic Bulletin*, **1**(6), 5–12.

McGee, T.G. and Yeung, Y.-M., 1993, Urban futures for Pacific Asia: towards the 21st century, in Yeung, Y.-M. (ed.), *Pacific Asia in the 21st Century*, Hong Kong: Chinese University Press.

McIlwaine, C., 1995, Gender, race and ethnicity: concepts, realities and policy implications, *Third World Planning Review*, **17**(2), 237–243.

Mehmet, O., 1995, *Westernizing the Third World: The Eurocentricity of Economic Development Theories*, London: Routledge.

Mehta, S.K., 1964, Some demographic and economic correlates of primate cities: a case for revaluation, *Demography*, **1**, 136–147.

Meier, G.M. and Baldwin, R.E., 1957, *Economic Development: Theory, History, Policy*, New York: John Wiley.

Mellaart, J., 1964, A Neolithic city in Turkey, *Scientific American Reprint*, 620, 10 pp.

Mellaart, J., 1967, *Çatal Höyük: a Neolithic City in Anatolia*, Oxford: Oxford University Press.

Mera, K., 1973, On the urban agglomeration and economic efficiency, *Economic Development and Cultural Change*, **21**, 309–324.

Mera, K., 1975, *Income Distribution and Regional Development*, University of Tokyo Press.

Mera, K., 1978, The changing pattern of population distribution in Japan and its implications for developing countries, in Lo, F.C. and Salih, K. (eds), *Growth Pole Strategies and Regional Development Policy*, Oxford: Pergamon.

Middleton, N., O'Keefe, P. and Moyo, S., 1993, *Tears of a Crocodile: From Rio to Reality in the Developing World*, London: Pluto.

Mies, M., 1993, The myth of catching-up development, in Mies, M. and Shiva, V. (eds), *Ecofeminism*, London: Zed.

Mies, M. and Shiva, V., 1993, *Ecofeminism*, London: Zed.

Mies, M., Bennholdt-Thomsen, V. and von Werlhof, C., 1988, *Women, The Last Colony*, London: Zed.

Miliband, R., 1977, *Marxism and Politics*, Oxford: Oxford University Press.

Miller, D., 1992, The young and the restless in Trinidad: a case of the local and the global in mass consumption, in Silverstone, R. and Hirsch, E. (eds), *Consuming Technology*, London: Routledge, 163–182.

Miller, D., 1994, *Modernity: An Ethnographic Approach: Dualism and Mass Consumption in Trinidad*, Oxford: Berg.

Milner-Smith, R. and Potter, R.B., 1995, Public knowledge of, and attitudes towards, the Third World, *Centre for Developing Areas Research (CEDAR) Research Papers*, 13, 32 pp.

Mitter, S., 1994, On organising women in casualised work: a global overview, in Rowbotham, S. and Mitter, S. (eds), *Dignity and Daily Bread: New Forms of Economic Organising among Poor Women in the Third World and the First*, London: Routledge.

Mohanty, C., 1991, Introduction: cartographies of struggle: Third World women and the politics of feminism, in Mohanty, C., Russo, A. and Torres, L. (eds), *Third World Women and the Politics of Feminism*, Bloomington: Indiana University Press.

Momsen, J. (ed.), 1993, *Women and Change in the Caribbean: A Pan-Caribbean Perspective*, London: James Currey.

Momsen, J. and Kinnard, V. (eds), 1992, *Different Places, Different Voices: Gender and Development in Africa, Asia and Latin America*, London: Routledge.

Morris, A.S., 1978, Urban growth patterns in Latin America with illustrations from Caracas, *Urban Studies*, **15**, 299–312.

Moseley, M.J., 1973, Growth centres – a shibboleth?, *Area*, **5**, 143–150.

Moseley, M.J., 1974, *Growth Centres in Spatial Planning*, Oxford: Pergamon.

Moser, C. and Peake, L. (eds), 1987, *Women, Human Settlements and Housing*, London: Tavistock.

Moser, C., Dennis, F. and Castleton, D., 1991, The urban context: women, settlements and the environment, in Sontheimer, S. (ed.), *Women and the Environment: A Reader*, London: Earthscan.

Moser, C.O.N., 1978, Informal sector or petty commodity production: dualism or dependence in urban development?, *World Development*, **6**, 1041–1064.

Moser, C.O.N., 1984, The informal sector reworked: viability and vulnerability in urban development, *Regional Development Dialogue*, **5**, 135–178.

Moser, C.O.N., 1989a, Gender planning in the Third World: meeting practical and strategic gender needs, *World Development*, **17**(11), 1799–825.

Moser, C.O.N., 1989b, The impact of reccession on structural adjustment at the micro-level low income women and households in Guayaquil, Ecuador, in UNICEF (ed.) *Invisible Adjustment*, **Vol II**, New York: UNICEF Americas Regional Office.

Moser, C.O.N., 1993, *Gender Planning and Development: Theory, Practice and Training*, London: Routledge.

Moser, C.O.N., 1995a, Urban social policy and poverty reduction, *Environment and Urbanisation*, **7**, 159–171.

Moser, C.O.N., 1995b, Women's mobilisation in human settlements: the case of Barrio Indio Guayas, in Kirby, J., O'Keefe, P. and Timberlake, L. (eds), *Sustainable Development: A Reader*, London: Earthscan.

Moser, C.O.N., 1996, Confronting crisis: a comparative study of household responses to poverty and vulnerability in four poor urban communities, *Environmentally Sustainable Development Studies and Monograph Series*, **8**, Washington, DC: World Bank.

Mountjoy, A.B., 1980, Worlds without end, *Third World Quarterly*, **2**, 753–757.

Mueller, C.C., 1995, Environmental problems inherent to a development style: degradation and poverty in Brazil, *Environment and Urbanisation*, **7**, 67–84.

Munasinghe, M., 1993, The economist's approach to sustainable development, *Finance and Development*, **30**(4), 16–19.

Murray, P. and Szelenyi, I., 1984, The city in the transition to socialism, *International Journal of Urban and Regional Research*, **8**, 90–107.

Myrdal, G., 1957, *Economic Theory and Underdeveloped Areas*, London: Duckworth.

Myrdal, G., 1968, *Asian Drama*, New York: Pantheon.

Nash, J. and Fernandez-Kelly, P. (eds), 1983, *Women, Men, and the International Division of Labour*, New York: State University of New York Press.

Nieuwenhuys, O., 1994, *Children's Lifeworlds: Gender, Welfare and Labour in the Developing World*, London: Routledge.

Norgaard, R.B., 1994, *Development Betrayed: The End of Progress and a Coevolutionary Revisioning of the Future*, London: Routledge.

Oberai, A.S., 1993, *Population Growth, Employment and Poverty in Third World Mega-Cities: Analytical and Policy Issues*, Basingstoke and London: Macmillan; New York: St Martin's Press.

O'Connor, A., 1983, *The African City*, London: Hutchinson.

O'Connor, A., 1991, *Poverty in Africa: A Geographical Approach*, London: Belhaven.

O'Connor, T., 1976, 'Third World' or one World?, *Area*, **8**, 269–271.

Okpala, D.C.I., 1996, Viewpoint: the second United Nations Conference on human settlements (Habitat II), *Third World Planning Review*, **18**, iii-xii.

Østergaard, L. (ed.), 1992, *Gender and Development: A Practical Guide*, London: Routledge.

Pahl, R.E., 1988, *On Work*, Oxford: Basil Blackwell.

Painter, J., 1987, *Guatemala: False Hope, False Freedom*, London: Latin American Bureau.

Pannell, C.W., 1990, China's urban geography, *Progress in Human Geography*, **14**, 214–236.

Parr, J.B., 1973, Growth poles, regional development, and central place theory, *Papers of the Regional Science Association*, **31**, 173–212.

Patullo, P., 1996, *Last Resort? Tourism in the Caribbean*, London: Mansell and the Latin American Bureau.

Payne, G.R. (ed.), 1984, *Low-Income Housing in the Developing World: the Role of Sites and Services and Settlement Upgrading*, Chichester: Wiley.

Peake, L., 1997, From cooperative socialism to a social housing policy? Declines and revivals in housing policy in Guyana, in Potter, R.B. and Conway, D. (eds), *Self-Help Housing, the Poor and the State in the Caribbean*, Knoxville: Tennessee University Press, 120–140.

Pearce, D., Markandya, A. and Barbier, E., 1989, *Blueprint for a Green Economy*, London: Earthscan.

Pearson, R., 1994, Gender relations, capitalism and Third World industrialisation, in Sklair, L. (ed.), *Capitalism and Development*, London: Routledge.

Pedersen, P.O., 1970, Innovation diffusion within and between national urban systems, *Geographical Analysis*, **2**, 203–254.

Peek, P. and Standing, G., 1982, *State Policies and Migration: Studies in Latin America and the Caribbean*, Beckenham: Croom Helm.

Peil, M., 1994, Urban housing and services in Anglophone West Africa, in Main, H. and Williams, S.W. (eds), *Environment and Housing in Third World Cities*, Chichester: Wiley.

Perlman, J.E., 1976, *The Myth of Marginality*, Berkeley: University of California Press.

Perloff, H.S. and Wingo, L., 1961, Natural resource endowment and regional economic growth, in Spengler, J.J. (ed.), *Natural Resources and Economic Growth*, Washington, DC: Resources for the Future.

Perroux, F., 1950, Economic space: theory and applications, *Quarterly Journal of Economics*, **64**, 89–104.

Perroux, F., 1955, Note sur la notion de 'pâte de croissance', *Economie Appliquée*, **1–2**, 307–320.

Phantumvanit, D. and Panayotou, T., 1990, *Industrialisation and Environmental Quality: Paying the Price*, Bangkok: Thailand Development Research Institute.

Phillip, P., 1989, *Urban low-income housing in St Lucia: an analysis of the formal and informal sectors*, unpublished MPhil thesis, University of London.

Phillips, D., 1990, *Health and Health Care in the Third World*, Harlow: Longman.

Phillips, D. and Verhasselt, Y. (eds), 1994, *Health and Development*, London: Routledge.

Polanyi, K., 1968, *Primitive, Archaic and Modern Economies: Essays of Karl Polanyi*, Boston.

Portes, A. and Walton, J., 1981, *Labor, Class and the International System*, New York: Academic Press.

Portes, A., Castells, M. and Benton, L. (eds), 1991, *The Informal Economy: Studies in Advanced and Less Developed Countries*, London: Johns Hopkins.

Potter, R.B., 1981, Industrial development and urban planning in Barbados, *Geography*, **66**, 225–228.

Potter, R.B., 1983, Tourism and development: the case of Barbados, *Geography*, **68**, 46–50.

Potter, R.B., 1984, Spatial perceptions and public involvement in Third World urban planning: the example of Barbados, *Singapore Journal of Tropical Geography*, **5**, 30–44.

Potter, R.B., 1985, *Urbanisation and Planning in the Third World: Spatial Perceptions and Public Participation*, Beckenham: Croom Helm; New York: St Martin's Press.

Potter, R.B., 1986a, Housing upgrading in Barbados, West Indies: the Tenantries Programme since 1980, *Geography*, **71**, 255–257.

Potter, R.B., 1986b, Spatial inequalities in Barbados, *Transactions of the Institute of British Geographers*, **11**, 183–198.

Potter, R.B., 1989a, Urban housing in Barbados, West Indies, *Geographical Journal*, **155**, 81–93.

Potter, R.B. (ed.), 1989b, *Urbanization, Planning and Development in the Caribbean*, London and New York: Mansell.

Potter, R.B., 1990, Cities, convergence, divergence and Third World development, Chapter 1 in Potter, R.B. and Salau, A.T. (eds), *Cities and Development in the Third World*, London and New York: Mansell.

Potter, R.B., 1992a, *Housing Conditions in Barbados: a Geographical Analysis*, Mona, Kingston, Jamaica: Institute of Social & Economic Research, University of the West Indies.

Potter, R.B., 1992b, *Urbanisation in the Third World*, Oxford: Oxford University Press.

Potter, R.B., 1993a, Basic needs and development in the small island states of the Eastern Caribbean, in Lockhart, D. and Drakakis-Smith, D. (eds), *Small Island Development*, London and New York: Routledge.

Potter, R.B., 1993b, Urbanization in the Caribbean and trends of global convergence– divergence, *Geographical Journal*, **159**, 1–21.

Potter, R.B., 1994, *Low-Income Housing and the State in the Eastern Caribbean*, Barbados, Jamaica, Trinidad and Tobago: The University Press of the West Indies.

Potter, R.B., 1995, Urbanisation and development in the Caribbean, *Geography*, **80**, 334–341.

Potter, R.B., 1996, Environmental impacts of urban-industrial development in the tropics: an overview, in Eden, M. and Parry, J.T. (eds), *Land Degradation in the Tropics*, London: Pinter.

Potter, R.B., 1997, Third World urbanisation in a global context, *Geography Review*, **10**, 2–6.

Potter, R.B. and Binns, J.A., 1988, Power, politics and society, in Pacione, M. (ed.), *The Geography of the Third World: Progress and Prospects*, London and New York: Routledge, 271–310.

Potter, R.B. and Conway, D. (eds), 1997, *Self-Help Housing, the Poor and the State in the Caribbean*, Knoxville: Tennessee University Press; Barbados, Jamaica and Trinidad and Tobago: The University Press of the West Indies.

Potter, R.B. and Dann, G.M.S., 1990, Dependent urbanization and retail change in Barbados, West Indies, in Potter, R.B. and Salau, A.T. (eds), *Cities and Development in the Third World*, London and New York: Mansell.

Potter, R.B. and Dann, G.M.S., 1994, Some observations concerning postmodernity and sustainable development in the Caribbean, *Caribbean Geography*, **5**, 92–107.

Potter, R.B. and Dann, G., 1996, Globalization, postmodernity and development in the Commonwealth Caribbean, in Yeung, Y.-M. (ed.), *Global Change and the Commonwealth*, Hong Kong Institute of Asia-Pacific Studies, Chinese University of Hong Kong, 103–129.

Potter, R.B. and Hunte, M., 1979, Recent developments in planning the settlement hierarchy of Barbados: implications concerning the debate on urban primacy, *Geoforum*, **10**, 355–362.

Potter, R.B. and O'Flaherty, P., 1995, An analysis of housing conditions in Trinidad and Tobago, *Social and Economic Studies*, **44**, 165–183.

Potter, R.B. and Salau, A.T. (eds), 1990, *Cities and Development in the Third World*, London and New York: Mansell.

Potter, R.B. and Unwin, T. (eds), 1989, *The Geography of Urban–Rural Interaction in Developing Countries*, London and New York: Routledge.

Potter, R.B. and Unwin, T., 1995, Urban–rural interaction: physical form and political process in the Third World, *Cities*, **12**, 67–73.

Potter, R.B. and Welch, B., 1996, Indigenization and development in the Caribbean, *Caribbean Week*, **8**, 13–14.

Potter, R.B. and Wilson, M.G., 1990, Indigenous environmental learning in a small developing country: adolescents in Barbados, West Indies, *Singapore Journal of Tropical Geography*, **11**, 56–67.

Pounds, N.J.G., 1969, The urbanization of the Classical World, *Annals of the Association of American Geographers*, **59**, 135–157.

Prasad, R.M. and Furedy, C., 1992, Small businesses from urban wastes: shoe renovation in Delhi, *Environment and Urbanisation*, **4**(2), 59–61.

Prebish, R., 1950, *The Economic Development of Latin America*, New York: United Nations.

Pred, A., 1977, *City-Systems in Advanced Economies*, London: Hutchinson.

Pred, A.R., 1973, The growth and development of systems of cities in advanced economies, in Pred, A. and Törnqvist, G. (eds), *Systems of Cities and Information Flows: Two Essays*, Lund Studies in Geography, Series B, **38**, 9–82.

Pryer, J. and Crook, N., 1988, *Cities of Hunger: Urban Malnutrition in Developing Countries*, Oxford: Oxfam.

Pugh, C., 1995, Urbanization in developing countries: an overview of the economic and policy issues in the 1990s, *Cities*, **6**, 381–398.

Putnam, R.D., 1993, *Making Democracy Work: Civic Traditions in Modern Italy*, Princeton, NJ: Princeton University Press.

Quijano, A., 1974, The marginal pole of the economy and the marginalised labour force, *Economy and Society*, **3**, 393–428.

Rabinovitch, J., 1992, Curitiba: towards sustainable development, *Environment and Urbanisation*, **4**(2), 62–73.

Radcliffe, S., 1996, Race, gender and generation: cultural geographies, in Preston, D. (ed.), *Latin American Development: Geographical Perspectives*, Harlow: Longman.

Radcliffe, S. and Westwood, S., 1993, *Viva: Women and Popular Protest in Latin America*, London: Routledge.

Rakodi, C., 1985, Self-reliance or survival? Food production in African cities, with particular reference to Lusaka, *African Urban Studies*, **21**, 53–63.

Ramirez, R., Fiori, J., Harms, H. and Mathéy, K., 1992, The commodification of self-help housing and state intervention: housing experiences in the Barrios of Caracas, in Mathéy, K. (ed.), *Beyond Self-Help Housing*, London and New York: Mansell, 95–144.

Rathgeber, E.M., 1990, WID, WAD, GAD: Trends in research and practice, *Journal of Developing Areas*, **24**, 489–502.

Redclift, M., 1987, *Sustainable Development: Exploring the Contradictions*, London: Routledge.

Redclift, M., 1992, The meaning of sustainable development, *Geoforum*, **23**(3), 395–403.

Redclift, M., 1994a, Development and the environment: managing the contradictions, in Sklair, L. (ed.), *Capitalism and Development*, London: Routledge.

Redclift, M., 1994b, Reflections on 'sustainable development' debate, *International Journal of Sustainable Development: World Ecology*, **1**, 3–21.

Redclift, M. and Sage, C. (eds), 1994, *Strategies for Sustainable Development: Local Agendas for the Southern Hemisphere*, Chichester: John Wiley.

Redclift, N. and Sinclair, M.T. (eds), 1991, *Working Women: International Perspectives on Labour and Gender Ideology*, London: Routledge.

Rees, W.E., 1990, The ecology of sustainable development, *The Ecologist*, **20**(1), 18–23.

Renaud, B., 1981, *National Urbanization Policy in Developing Countries*, Oxford: Oxford University Press for the World Bank.

Reynolds, L.G., 1969, Economic development with surplus labour: some complications, *Oxford Economic Papers*, **21**, 89–103.

Richardson, H.W., 1973a, Theory of the distribution of city sizes: review and prospects, *Regional Studies*, **7**, 239–251.

Richardson, H.W., 1973b, *The Economics of Urban Size*, Farnborough: Saxon House.

Richardson, H.W., 1976, The argument for very large cities reconsidered: a comment, *Urban Studies*, **13**, 307–310.

Richardson, H.W., 1977, City size and national spatial strategies in developing countries, *World Bank Staff Working Paper*, 252.

Richardson, H.W., 1978, *Regional and Urban Economics*, Harmondsworth: Penguin.

Richardson, H.W., 1980, Polarization reversal in developing countries, *Papers of the Regional Science Association*, **45**, 67–85.

Richardson, H.W., 1981, National urban development strategies in developing countries, *Urban Studies*, **18**, 267–283.

Richardson, H.W., 1989, The big, bad city: mega-city myth?, *Third World Planning Review*, **11**, 355–372.

Richardson, H.W. and Richardson, M., 1975, The relevance of growth center strategies to Latin America, *Economic Geography*, **51**, 163–178.

Riddell, J.B., 1970, *The Spatial Dynamics of Modernization in Sierra Leone: Structure, Diffusion and Response*, Illinois: Evanston.

Roberts, B., 1978, *Cities of Peasants: the Political Economy of Urbanization in the Third World*, London: Arnold.

Roberts, B., 1995, *The Making of Citizens: Cities of Peasants Revisited*, London: Arnold.

Roberts, B.R., 1991, Employment structure, life cycle, and life chances: formal and informal sectors in Guadalajara, in Portes, A., Castells, M. and Benton, L. (eds), *The Informal Economy: Studies in Advanced and Less Developed Countries*, London: Johns Hopkins.

Robertson, A., 1994, Free trade or fair trade, *Anti-Slavery International Reporter*, **13**(9), 63–64.

Robins, K., 1989, Global times, *Marxism Today*, December, 20–27.

Robins, K., 1995, The new spaces of global media, in Knox, P.C. and Taylor, P.J. (eds), *World Cities in a World-System*, Cambridge: Cambridge University Press, 248–262.

Robins, N. and Trisoglio, 1995, Restructuring industry for sustainable development, in Kirby, J., O'Keefe, P. and Timberlake, L. (eds), *Sustainable Development: A Reader*, London: Earthscan.

Robson, B.T., 1973, *Urban Growth: An Approach*, London: Methuen.

Rogers, B., 1980, *The Domestication of Women: Discrimination in Developing Societies*, London: Tavistock.

Rogerson, C.W. and Hart, D.M., 1989, The survival of the 'informal sector': the shebeens of black Johannesburg, *GeoJournal*, **12**(2), 153–166.

Rojas, E., 1989, Human settlements of the Eastern Caribbean: development problems and policy options, *Cities*, **6**, 243–258.

Rojas, E., 1995, Commentary: Government–market interactions in urban development policy, *Cities*, **12**, 399–400.

Rondinelli, D.A., 1982, Intermediate cities in developing countries: a comparative analysis of their demographic, social and economic characteristics, *Third World Planning Review*, **4**, 357–386.

Rondinelli, D.A., 1983a, Dynamics of growth of secondary cities in developing countries, *Geographical Review*, **73**, 42–57.

Rondinelli, D.A., 1983b, *Secondary Cities in Developing Countries: Policies for Diffusing Urbanization*, London: Sage.

Rostow, W.W., 1960, *The Stages of Economic Growth: A Non-Communist Manifesto*, London: Cambridge University Press.

Sabot, R.H., 1979, *Economic Development and Urban Migration in Tanzania 1900–1971*, Oxford: Clarendon Press.

Safa, H., 1981, Runaway shops and female employment: the search for cheap labour, *Signs*, **7**(2), 418–433.

Safa, H., 1986, Economic autonomy and sexual equality in Caribbean society, *Social and Economic Studies*, **35**(3), 1–22.

Safa, H., 1990, Women and industrialisation in the Caribbean, in Stitcher, S. and Parpart, J. (eds), *Women, Employment and the Family in the International Division of Labour*, London: Macmillan.

Safier, M., 1969, Towards the definition of patterns in the distribution of economic development over East Africa, *East African Geographical Review*, **7**, 1–13.

Sandbrook, R., 1985, *The Politics of Africa's Economic Stagnation*, Cambridge: Cambridge University Press.

Santos, M., 1979, *The Shared Space: The Two Circuits of the Urban Economy in Underdeveloped Countries*, London: Methuen.

Sanyal, B., 1986, *Urban Cultivation in East Africa, Food and Energy Nexus Programme*, Tokyo: United Nations University.

Sassen, S., 1988, *The Mobility of Labour and Capital*, Cambridge: Cambridge University Press.

Sassen, S., 1991, *The Global City*, Princeton, NJ: Princeton University Press.

Sassen, S., 1996, The global city, in Fainstein, S. and Campbell, S. (eds), *Readings in Urban Theory*, Oxford: Blackwell, 61–71.

Satterthwaite, D., 1992, Introduction: sustainable cities, *Environment and Urbanisation*, **4**(2), 3–8.

Satterthwaite, D., 1995, Viewpoint – The underestimation of urban poverty and of its health consequences, *Third World Planning Review*, **17**, iii–xii.

Sattaur, O., 1993, *Child Labour in Nepal*, London: Anti Slavery International.

Scarpaci, J.L. and Irarrazaval, I., 1996, Caring for people: health care and education provision, in Preston, D. (ed.), *Latin American Development: Geographical Perspectives*, Harlow: Longman.

Schenk, H., 1974, Concepts behind urban and regional planning in China, *Tijdschrift voor Economische en Sociale Geografie*, **65**, 381–389.

Schultz, T.W., 1953, *The Economic Organization of Agriculture*, New York: McGraw-Hill.

Schumpeter, J.A., 1911, *Die Theorie des Wirtschaftlichen Entwicklung*, Leipzig.

Schwarz, A., 1993, Looking back at Rio, *Far Eastern Economic Review*, **28**, 48–52.

Scott, A.M., 1979, Who are the self-employed?, in Bromley, R. and Gerry, C. (eds), *Casual Work and Poverty in Third World Cities*, Chichester: John Wiley.

Scott, A.M., 1986a, Industrialisation, gender segregation and stratification theory, in Crompton, R. and Mann, M. (eds), *Gender and Stratification*, Cambridge: Polity.

Scott, A.M., 1986b, Women and industrialisation: examining the 'female marginalisation' thesis, *Journal of Development Studies*, **22**(4), 649–680.

Scott, A.M., 1994, *Divisions and Solidarities: Gender, Class and Employment in Latin America*, London: Routledge.

Scott, A.M., 1995, Informal sector or female sector? Gender bias in labour market models, in Elson, D. (ed.), *Male Bias in the Development Process*, Manchester: Manchester University Press.

Seager, J. and Olson, A., 1986, *Women in the World: An International Atlas*, London: Pan.

Sen, A.K., 1982, *Poverty and Famines: An Essay on Entitlement and Deprivation*, Oxford: Oxford University Press.

Sen, G., 1988, Alternative visions, strategies and methods, in Sen, G. and Grown, C. (eds), *Development, Crises and Alternative Visions*, New York: MR Press.

Sen, G. and Grown, K., 1988, *Development, Crisis and Alternative Visions*, London: Earthscan.

Sethuraman, S.V., 1976, The urban informal sector: concept, measurement and policy, *International Labour Review*, **114**, 69–82.

Sethuraman, S.V., 1978, The informal urban sector in developing countries: some policy implications, in de Souza, A. (ed.), *The Indian City*, New Delhi: Manohor.

Sethuraman, S.V., 1981, *The Urban Informal Sector in Developing Countries: Employment, Poverty and Environment*, Geneva: International Labour Office.

Shiva, V., 1988, *Staying Alive: Women, Ecology and Survival in India*, London: Zed.

Shiva, V., 1992, Recovering the real meaning of sustainability, in Cooper, D.E. and Palmer, J.A. (eds), *The Environment in Question*, London: Routledge.

Sidaway, J.D., 1990, Post-Fordism, post-modernity and the Third World, *Area*, **22**, 301–303.

Simai, M. (ed.), 1995, *Global Employment: An International Investigation into the Future of Work, Volumes I and II*, London: Zed.

Simon, D., 1989a, Colonial cities: post-colonial Africa and the world economy: a reinterpretation, *International Journal of Urban and Regional Research*, **13**(1), 68–91.

Simon, D., 1989b, Sustainable development: theoretical construct or attainable goal?, *Environmental Conservation*, **16**, 41–48.

Simon, D. (ed.), 1990, *Third World Regional Development: A Reappraisal*, London: Paul Chapman.

Simon, D., 1992, *Cities, Capital and Development: African Cities in the World Economy*, London: Belhaven.

Simon, D., 1993, The world city hypothesis: reflections from the periphery, *Centre for Developing Areas Research Paper*, No. 7, Royal Holloway, University of London.

Simon, D., 1995, Commentary: Third World urbanization in the 1990s: a commentary on the views of Cedric Pugh, *Cities*, **12**, 401–403.

Simon, D., 1996, *Transport and Development in the Third World*, London: Routledge.

Simon, D. and Birch, S.L., 1992, Formalising the informal sector in a changing South Africa: small-scale manufacturing in the Witwatersrand, *World Development*, **20**(7), 1029–1045.

Simon, D., Van Spengen, W., Dixon, C. and Narman, A. (eds), 1995, *Structurally Adjusted Africa: Poverty, Debt and Basic Needs*, London: Pluto.

Simon, J., 1981, *The Ultimate Resource*, Princeton, NJ: Princeton University Press.

Simpson, G.E., 1954, Begging in Kingston and Montego Bay, *Social and Economic Studies*, **3**(2), 197–211.

Singer, H.W., 1970, Dualism revisited: a new approach to the problems of the dual society in developing countries, *Journal of Development Studies*, **7**, 60–75.

Sjoberg, G., 1960, *The Preindustrial City: Past and Present*, New York: Free Press.

Sjoberg, G., 1965, The origin and evolution of cities, *Scientific American*, September, reprinted as *Cities: Their Origin, Growth and Human Impact*, San Francisco: Freeman.

Skinner, R.J. and Rodell, M.J., 1983, *People, Poverty and Shelter: Problems of Self-Help Housing in the Third World*, London: Methuen.

Sklair, L., 1991, *Sociology of the Global System*, Hemel Hempstead: Harvester Wheatsheaf.

Sklair, L., 1994a, Global system, local problems: environmental impacts of transnational corporations along Mexico's northern border, in Main, H. and Williams, S.W. (eds), *Environment and Housing in Third World Cities*, Chichester: Wiley.

Sklair, L. (ed.), 1994b, *Capitalism and Development*, London: Routledge.

Smith, D.M. (ed.), 1986, *Living under Apartheid*, London: Allen and Unwin.

Smith, D.M., 1992, Redistribution after apartheid: who gets what where in the new South Africa, *Area*, **24**(4), 350–358.

Smith, D.M., 1995, Redistribution and social justice after apartheid, in Lemon, A. (ed.), *The Geography of Change in South Africa*, Chichester: Wiley.

Smyke, P., 1991, *Women and Health*, London: Zed.

Smyth, C.G., 1996, *The environmental impacts of urban land use change since 1965, Niteroi Municipality, Rio de Janeiro State, Brazil*, unpublished PhD Thesis, Queen's University, Belfast.

Soja, E.W., 1968, *The Geography of Modernization in Kenya: A Spatial Analysis of Social, Economic and Political Change*, Syracuse, NY: Syracuse University Press.

Soja, E.W., 1974, The geography of modernization: paths, patterns, and processes of spatial change in developing countries, in Bruner, R. and Brewer, G. (eds), *A Policy Approach to the Study of Political Development and Change*, New York: Free Press.

Sontheimer, S. (ed.), 1991, *Women and the Environment: A Reader*, London: Earthscan.

Standing, G., 1989, Global feminisation through flexible labour, *World Development*, **17**, 1077–1095.

Stephens, C. and Harpham, T., 1992, Health and the environment in urban areas of developing countries, *Third World Planning Review*, **14**(3), 2647–2682.

Stewart, F., 1985, *Planning to Meet Basic Needs*, London: Macmillan.

Stewart, F., 1995, *Adjustment and Poverty: Options and Choices*, Oxford: Blackwell.

Stitcher, S. and Parpart, J. (eds), 1990, *Women, Employment and the International Division of Labour*, London: Macmillan.

Stöhr, W.B., 1981, Development from below: the bottom-up and periphery-inward development paradigm, Chapter 2 in Stöhr, W.B. and Taylor, D.R.F. (eds), *Development from Above or Below? The Dialectics of Regional Planning in Developing Countries*, Chichester: Wiley, 39–72.

Stöhr, W.B. and Taylor, D.R.F., 1981, *Development from Above or Below? The Dialectics of Regional Planning in Developing Countries*, Chichester: Wiley.

Stokes, C., 1962, A theory of slums, *Land Economics*, **38**, 187–197.

Stren, R., White, R. and Whitney, J. (eds), 1992, *Sustainable Cities: Urbanization and the Environment in International Perspective*, Boulder, CO: Westview Press.

Stretton, H., 1978, *Urban Planning in Rich and Poor Countries*, Oxford: Oxford University Press.

Susman, P., 1987, Spatial equality and socialist transformation in Cuba, in Forbes, D. and Thrift, N. (eds), *The Socialist Third World: Urban Development and Territorial Planning*, Oxford: Blackwell.

Sweetman, C. (ed.), 1994, *North–South Co-operation*, Oxford: Oxfam.

Swyngedouw, E.A., 1995, The contradictions of urban water provision: a study of Guayaquil, Ecuador, *Third World Planning Review*, **17**, 387–405.

Taaffe, E.J., Morrill, R.L. and Gould, P.R., 1963, Transport expansion in underdeveloped countries: a comparative analysis, *Geographical Review*, **53**, 503–529.

Tang, W.-S., 1993, *A Critical Review of the English Literature on Chinese Urbanisation*, Hong Kong Institute of Asia-Pacific Studies, Chinese University of Hong Kong.

Taylor, J.L. and Williams, D.G. (eds), 1982, *Urban Planning Practice in Developing Countries*, Oxford: Pergamon.

Taylor, P.J., 1986, The world-systems project, Chapter 12 in Johnston, R.J. and Taylor, P.J. (eds), *A World in Crisis? Geographical Perspectives*, Oxford and New York: Blackwell, 333–354.

Teltscher, S., 1994, Small trade and the world economy: informal vendors in Quito, Ecuador, *Economic Geography*, **70**(2), 167–187.

Thomas, G.A., 1991, The gentrification of paradise: St John's, Antigua, *Urban Geography*, **12**, 469–487.

Thomas, J.J., 1992, *Informal Economic Activity*, Hemel Hempstead: Harvester Wheatsheaf.

Thomas, J.J., 1995, *Surviving in the City? The Urban Informal Sector in Latin America*, London: Pluto.

Thrift, N., 1986, The geography of the international economic order, in Johnston, R.J. and Taylor, P.J. (eds), *A World in Crisis*, Oxford: Blackwell.

Timberlake, L. and Thompson, L., 1990, *When the Bough Breaks: Our Children, Our Environment*, London: Earthscan.

Todaro, M.P., 1995, *Economic Development in the Third World*, Harlow: Longman.

Tokman, V., 1989, Policies for a heterogeneous informal sector in Latin America, *World Development*, **17**(7), 1067–1076.

Turner, A. (ed.), 1980, *The Cities of the Poor: Settlement Planning in Developing Countries*, London: Croom Helm.

Turner, A. and Smulian, J., 1971, New cities in Venezuela, *Town Planning Review*, **42**, 3–18.

Turner, J.F.C., 1963, Dwelling resources in South America, *Architectural Design*, **37**, 360–393.

Turner, J.F.C., 1967, Barriers and channels for housing development in modernizing countries, *Journal of the American Institute of Planners*, **33**, 167–181.

Turner, J.F.C., 1968a, Housing priorities, settlement patterns and urban development in modernizing countries, *Journal of the American Institute of Planners*, **34**, 354–363.

Turner, J.F.C., 1968b, The squatter settlement: an architecture that works, *Architectural Design*, **38**, 355–360.

Turner, J.F.C., 1969, Uncontrolled urban settlement: problems and policies, in Breese, G. (ed.), *The City in Newly Developing Countries*, Prentice-Hall, 507–535.

Turner, J.F.C., 1972, Housing as a verb, in Turner, J.F.C. and Fichter, R. (eds), *Freedom to Build: Dweller Control of the Housing Process*, London: Collier-Macmillan, 148–175.

Turner, J.F.C., 1976, *Housing by People: Towards Autonomy in Building Environments*, London: Marion Boyars.

Turner, J.F.C., 1982, Issues in self-help and self-managed housing, Chapter 4 in Ward, P.M. (ed.), *Self-Help Housing: A Critique*, London: Mansell, 99–113.

Turner, J.F.C., 1983, From central provision to local enablement: new directions for housing policies, *Habitat International*, **7**, 207–210.

Turner, J.F.C., 1985, Future directions in housing policies, paper delivered at the International Symposium on the Implementation of a Support Policy for Housing Provision, Development Planning Unit, University College London.

Turner, J.F.C., 1988, Issues and conclusions, in Turner, B. (ed.), *Building Community: A Third World Case Book*, London: Building Community Books.

Turner, J.F.C., 1990, Barriers, channels and community control, in Cadman, D. and Payne, G. (eds), *The Living City*, London: Routledge, 181–191.

Turnham, D., 1993, *Employment and Development: A New Review of Evidence*, Geneva: International Labour Office Publications.

Turnham, D., Salome, B. and Schwarz, A., 1990, *The Informal Sector Revisited*, Paris: OECD.

Udall, A.T. and Sinclair, S., 1982, The luxury unemployment hypothesis: a review of recent evidence, *World Development*, **10**, 49–62.

UNCED, 1992, The Earth Summit, *CONNECT* (UNESCO–UNEP Environmental Education Newsletter), **XVII**(2), 1–8.

UNICEF (United Nations Children's Fund), 1995, *The State of the World's Children 1995*, Oxford: Oxford University Press.

United Nations, 1973, *Urban Land Policies and Land-Use Control Measures, Vol. 1, Africa*, New York: United Nations.

United Nations, 1980, *Patterns of Urban and Rural Population Growth*, New York: United Nations.

United Nations, 1988, *World Population Trends and Policies: 1987 Monitoring Report*, New York: United Nations Department of International Economic and Social Affairs.

United Nations, 1989a, *Prospects for World Urbanization 1988*, New York: United Nations.

United Nations, 1989b, *Report on the World Social Situation*, New York: United Nations.

United Nations, 1991, *World Urbanisation Prospects 1990*, New York: United Nations.

United Nations, 1995, Preparatory committee for the United Nations Conference on Human Settlements (Habitat II): draft statement of principles and commitments and global plan of action, *The Habitat Agenda*, October.

United Nations Centre for Human Settlements (Habitat), 1996, *An Urbanizing World: Global Report on Human Settlements 1996*, Oxford and New York: Oxford University Press for the United Nations Centre for Human Settlements.

United Nations Development Programme, 1991, *Cities, People and Poverty: Urban Development Cooperation for the 1990s*, New York: United Nations Development Programme.

United Nations Development Programme, 1992, *Human Development Report 1992*, Oxford: Oxford University Press.

United Nations Development Programme, 1993, *Human Development Report 1993*, Geneva.

United Nations Development Programme, 1995, *Human Development Report 1995*, Oxford: Oxford University Press.

United Nations Development Programme, 1996, *Human Development Report 1996*, Oxford: Oxford University Press.

Urban Foundation, 1993, *Managing Urban Poverty*, Johannesburg: Urban Foundation.

Urry, J., 1990a, The consumption of tourism, *Sociology*, **24**, 23–35.

Urry, J., 1990b, *The Tourist Gaze*, London: Sage.

Vance, J., 1970, *The Merchant's World: The Geography of Wholesaling*, Englewood Cliffs, NJ: Prentice-Hall.

Vapnarsky, C.A., 1969, On rank-size distributions of cities: an ecological approach, *Economic Development and Cultural Change*, **17**, 584–595.

Vendovato, C., 1986, *Politics, Foreign Trade and Economic Development: A Study of the Dominican Republic*, Beckenham: Croom Helm.

von Böventer, E., 1975, Regional growth theory, *Urban Studies*, **12**, 1–29.

Walker, G., 1994, Industrial hazards, vulnerability and planning in Third World cities, with reference to Bhopal and Mexico City, in Main, H. and Williams, S.W. (eds), *Environment and Housing in Third World Cities*, Chichester: Wiley.

Wallace, T. and March, C., 1991, *Changing Perceptions*, Oxford: Oxfam.

Wallerstein, I., 1974, *The Modern World-System*, New York: Academic Press.

Wallerstein, I., 1980, *The Modern World System II: Mercantilism and the Consolidation of the European World Economy, 1600–1750*, New York: Academic Press.

Ward, B., 1980, First, second, third and fourth worlds, *Economist*, 18 May, 65–73.

Ward, P., 1976, The squatter settlement as slum or housing solution: some evidence from Mexico City, *Land Economics*, **52**, 330–346.

Ward, P.M. (ed.), 1982, *Self-Help Housing: A Critique*, London: Mansell.

Ward, P.M., 1986, *Welfare Politics in Mexico: Papering over the Cracks*, London: Unwin Hyman.

Ward, P.M., 1990, *Mexico City: The Production and Reproduction of an Urban Environment*, London: Belhaven.

Ward, P.M., 1993, The Latin American inner city: difference of degree or kind?, *Environment and Planning A*, **25**, 1131–1160.

Ward, P.M. and Macoloo, C., 1992, Articulation theory and self-help housing practice in the 1990s, *International Journal of Urban and Regional Research*, **6**, 60–80.

Watson, M. and Potter, R.B., 1993, Housing and housing policy in Barbados: the relevance of the chattel house, *Third World Planning Review*, **15**, 373–395.

Watson, M. and Potter, R.B., 1997, Housing conditions, vernacular architecture and state housing policy in Barbados, in Potter, R.B. and Conway, D. (eds), *Self-Help Housing, the Poor and the State in the Caribbean*, Knoxville: Tennessee University Press, 30–51.

Watson, S. and Gibson, K. (eds), 1995, *Postmodern Cities and Spaces*, Oxford (UK) and Cambridge (USA): Blackwell.

Weeks, J., 1975, Policies for expanding employment in the informal urban sector of developing countries, *International Labour Review*, **111**, 1–13.

Weir, D., 1995, Run into the wind, in Kirby, J., O'Keefe, P. and Timberlake, L. (eds), *Sustainable Development: A Reader*, London: Earthscan.

Wen, Y.-K. and Sengupta, J. (eds), 1991, *Increasing the International Competitiveness of Exports from Caribbean Countries*, Washington, DC: World Bank.

Westwood, S. and Radcliffe, S., 1993, Gender, racism and the politics of identities in Latin America, in Radcliffe, S. and Westwood, S. (eds), *Viva: Women and Popular Protest in Latin America*, London and New York: Routledge.

Wheatley, P., 1971, *The Pivot of the Four Quarters*, Chicago: Chicago University Press.

White, R. and Whitney, J., 1992, Cities and the environment: an overview, in Stren, R., White, R. and Whitney, J. (eds), *Sustainable Cities*, Boulder, CO: Westview Press.

White, R.R., 1994, Strategic decisions for sustainable urban development in the Third World, *Third World Planning Review*, **16**, 103–116.

WHO Commission on Health and the Environment, 1992, Our planet, our health, *Environment and Urbanisation*, **4**(1), 65–76; *Our Planet, Our Health*, Report of the Commission on Health and Environment, Geneva: WHO Publications.

Williams, D.G., 1984, The role of international agencies: the World Bank, Chapter 10 in Payne, G.K. (ed.), *Low-Income Housing in the Developing World*, Chichester: Wiley.

Williamson, J.G., 1965, Regional inequality and the process of national development: a description of the patterns, *Economic Development and Cultural Change*, **13**, 3–45.

Wolf-Phillips, L., 1979, Why Third World?, *Third World Quarterly*, **1**, 105–113.

Wolpe, H. (ed.), 1980, *The Articulation of Modes of Production*, London: Routledge and Kegan Paul.

World Bank, 1986, *Population Growth and Policies in Sub-Saharan Africa*, Washington, DC: World Bank.

World Bank, 1990, *World Development Report*, Oxford: Oxford University Press.

World Bank, 1991, *Urban Policy and Economic Development: An Agenda for the 1990s*, Washington, DC: World Bank.

World Bank, 1993a, *Towards Environmental Strategies for Cities*, Urban Development Division Strategy Paper, Washington, DC.

World Bank, 1993b, *World Development Report 1993: Investing in Health*, Oxford: Oxford University Press.

World Bank, 1996, *World Bank Development Report*.

World Commission on Environment and Development (WCED), 1987, *Our Common Future*, Oxford: Oxford University Press.

World Conservation Strategy, 1991, *Caring for the Earth*, Geneva: IUCN-UNEP-WWF.

Wratten, E., 1995, Conceptualising urban poverty, *Environment and Urbanisation*, **7**, 11–36.

Wu, C.T. and Ip, D.F., 1981, China: rural development – alternating combinations of top-down and bottom-up strategies, in Stöhr, W.B. and Taylor, D.R.F. (eds), *Development from Above or Below?*, Chichester: John Wiley.

Yeh, A.G.O., 1990, Unfair housing subsidy and public housing in Hong Kong, *Environment and Planning C: Government and Policy*, **8**, 439–454.

Yeh, A.G.O. and Fong, P.K.W., 1984, Public housing and urban development in Hong Kong, *Third World Planning Review*, **6**, 79–94.

Yelvington, K., 1993, *Trinidad Ethnicity*, London: Macmillan.

Yeung, Y.-M., 1990, *Changing Cities of Pacific Asia: A Scholarly Interpretation*, Hong Kong: Chinese University of Hong Kong.

Yeung, Y.-M., 1995, Commentary: urbanization and the NPE: an Asia-Pacific perspective, *Cities*, **12**, 409–411.

Young, K., 1993, *Planning Development with Women*, Basingstoke: Macmillan.

Zipf, G.K., 1949, *Human Behaviour and the Principle of Least Effort*, Cambridge, Massachusetts: MIT Press.

Index